Methods in Molecular Biology

Series Editor
John M. Walker
School of Life and Medical Sciences
University of Hertfordshire
Hatfield, Hertfordshire, UK

For further volumes:
http://www.springer.com/series/7651

For over 35 years, biological scientists have come to rely on the research protocols and methodologies in the critically acclaimed Methods in Molecular Biology series. The series was the first to introduce the step-by-step protocols approach that has become the standard in all biomedical protocol publishing. Each protocol is provided in readily-reproducible step-by-step fashion, opening with an introductory overview, a list of the materials and reagents needed to complete the experiment, and followed by a detailed procedure that is supported with a helpful notes section offering tips and tricks of the trade as well as troubleshooting advice. These hallmark features were introduced by series editor Dr. John Walker and constitute the key ingredient in each and every volume of the Methods in Molecular Biology series. Tested and trusted, comprehensive and reliable, all protocols from the series are indexed in PubMed.

Manuscript Instructions https://drive.google.com/file/d/1K4Az5vpAFVzYadJ3jKx2-QSo7Dqa_SNA/view?usp=sharing

Botrytis cinerea

Methods and Protocols

Edited by

Luca Nerva

Centre for Viticulture and Enology, CREA-VE, Conegliano, Treviso, Italy

Walter Chitarra

Centre for Viticulture and Enology, CREA-VE, Conegliano, Treviso, Italy

Humana Press

Editors
Luca Nerva
Centre for Viticulture and Enology
CREA-VE
Conegliano, Treviso, Italy

Walter Chitarra
Centre for Viticulture and Enology
CREA-VE
Conegliano, Treviso, Italy

ISSN 1064-3745 ISSN 1940-6029 (electronic)
Methods in Molecular Biology
ISBN 978-1-0716-5550-4 ISBN 978-1-0716-5551-1 (eBook)
https://doi.org/10.1007/978-1-0716-5551-1

This Humana imprint is published by the registered company Springer Science+Business Media, LLC, part of Springer Nature.
The registered company address is: 1 New York Plaza, New York, NY 10004, U.S.A.

Preface

Botrytis cinerea is a major necrotrophic fungal pathogen responsible for severe yield and postharvest losses worldwide. Its broad host range, genetic plasticity, and capacity to develop fungicide resistance make it a key model for studying fungal pathogenicity and host–pathogen interactions. This volume compiles validated protocols spanning classical and advanced methodologies, including isolation and culturing techniques, microscopy, inoculation assays, antifungal susceptibility testing, protoplast production, RNA interference, mycovirus detection, and omics-based approaches. By integrating traditional phytopathology with molecular and systems-level tools, this book provides a practical reference to promote reproducibility and innovation in *Botrytis* research.

Botrytis cinerea is among the most economically relevant necrotrophic fungal pathogens, affecting more than 200 plant species and causing substantial yield and postharvest losses worldwide. Beyond its agricultural importance, Botrytis has emerged as a model organism for investigating fundamental questions in fungal biology, including mechanisms of pathogenicity, host colonization strategies, stress adaptation, fungicide resistance, and genome evolution.

Over the last decades, research on Botrytis has evolved from classical phytopathological investigations to highly interdisciplinary studies integrating molecular genetics, cell biology, genomics, transcriptomics, proteomics, and metabolomics. The increasing availability of genome sequences, functional genomics tools, and high-resolution imaging approaches has greatly expanded our capacity to dissect the molecular dialogue between pathogen and host. At the same time, the diversification of experimental platforms has highlighted the urgent need for harmonized, reproducible, and well-documented methodologies that can ensure comparability across laboratories and experimental systems.

This volume was conceived as a comprehensive methodological reference dedicated specifically to *Botrytis* research. It brings together contributions from internationally recognized experts, providing detailed and robust protocols that span the entire experimental workflow. The chapters cover classical approaches, including isolation, cultural techniques, sclerotia induction, and storage, alongside advanced methodologies such as microscopy and protein localization, inoculation assays on fruits and vegetables, antifungal susceptibility testing, protoplast production and transfection, RNA interference, and identification of mycoviral infections. In addition, the volume presents state-of-the-art molecular and omics-based approaches, including high-quality DNA and RNA extraction, molecular marker development, genome assembly, quantitative gene expression analysis, and integrative transcriptomic, proteomic, and metabolomic analyses.

By combining traditional techniques with modern molecular tools, this book aims to serve both early-career researchers entering the field and experienced scientists seeking to expand their methodological repertoire. We hope that this collection will promote experimental rigor, facilitate cross-disciplinary collaboration, and stimulate innovative research directions in *Botrytis* biology, ultimately contributing to more sustainable and effective disease management strategies.

Conegliano, Italy — *Dr. Luca Nerva*
Conegliano, Italy — *Dr. Walter Chitarra*

Contents

Contributors

MARÍA A. AYLLÓN • *Centro de Biotecnología y Genómica de Plantas, Universidad Politécnica de Madrid (UPM)-Instituto Nacional de Investigación y Tecnología Agraria y Alimentaria (INIA), Campus de Montegancedo, Pozuelo de Alarcón, Madrid, Spain; Departamento de Biotecnología-Biología Vegetal, Escuela Técnica Superior de Ingeniería Agronómica, Alimentaria y de Biosistemas, Universidad Politécnica de Madrid (UPM), Madrid, Spain*

LOUIS BACKMANN • *Institute for Viticulture and Oenology, Dienstleistungszentrum Ländlicher Raum (DLR) Rheinpfalz, Neustadt, Germany*

MARC BARDIN • *INRAE, Pathologie Végétale, Montfavet, France*

JEAN-FRANÇOIS BOURGEAY • *INRAE, Pathologie Végétale, Montfavet, France*

MARIA VICTORIA CALCIS-MARZÁN • *Microbiology and Proteomics Laboratory, Institute for Viticulture and Agri-Food Research (IVAGRO), Faculty of Environmental and Marine Sciences, University of Cádiz,, Puerto Real, Spain*

DAOGUO CHEN • *Key Laboratory of Plant Resources, The Innovative Academy of Seed Design, Institute of Botany, Chinese Academy of Sciences, Beijing, China; China National Botanical Garden, Beijing, China; University of Chinese Academy of Sciences, Beijing, China*

TONG CHEN • *State Key Laboratory of Plant Diversity and Specialty Crops, Institute of Botany, Chinese Academy of Sciences, Beijing, China; China National Botanical Garden, Beijing, China*

YONG CHEN • *State Key Laboratory of Plant Diversity and Specialty Crops, Institute of Botany, Chinese Academy of Sciences, Beijing, China; China National Botanical Garden, Beijing, China;*

MARCO CHIAPELLO • *Institute for Sustainable Plant Protection, National Research Council of Italy, Turin, Italy*

LAURA CÓRDOBA • *Centro de Biotecnología y Genómica de Plantas, Universidad Politécnica de Madrid (UPM)-Instituto Nacional de Investigación y Tecnología Agraria y Alimentaria (INIA/CSIC), Campus de Montegancedo, Pozuelo de Alarcón, Madrid, Spain*

OLIVIER COSTE • *Microbiology and Proteomics Laboratory, Institute for Viticulture and Agri-Food Research (IVAGRO), Faculty of Environmental and Marine Sciences, University of Cádiz,, Puerto Real, Spain*

MARGHERITA D'AMICO • *Council for Agricultural Research and Economics–Research Center for Viticulture and Enology (CREA-VE), Via Casamassima, Turi BA, Italy*

CRUCITTI DALILA • *Institute of Biosciences and BioResources (IBBR), Italian National Research Council (CNR), Palermo, Italy*

PACIFICO DAVIDE • *Institute of Biosciences and BioResources (IBBR), Italian National Research Council (CNR), Palermo, Italy*

ALMUDENA ESCOBAR-NIÑO • *Microbiology and Proteomics Laboratory, Institute for Viticulture and Agri-Food Research (IVAGRO), Faculty of Environmental and Marine Sciences, University of Cádiz,, Puerto Real, Spain*

FRANCISCO JAVIER FERNANDEZ-ACERO • *Microbiology and Proteomics Laboratory, Institute for Viticulture and Agri-Food Research (IVAGRO), Faculty of Environmental and Marine Sciences, University of Cádiz,, Puerto Real, Spain*

ANDRÉ FLEISSNER • *Institute for Genetics, Technische Universität Braunschweig, Braunschweig, Germany*

LIDA HALILOVIC • *Department of Microbiology and Plant Pathology, Center for Plant Cell Biology, Institute for Integrative Genome Biology, University of California, Riverside, CA, USA*

PO HU • *Department of Microbiology and Plant Pathology, Center for Plant Cell Biology, Institute for Integrative Genome Biology, University of California, Riverside, CA, USA*

XINHUA HUANG • *State Key Laboratory of Plant Diversity and Specialty Crops, Institute of Botany, Chinese Academy of Sciences, Beijing, China; China National Botanical Garden, Beijing, China; University of Chinese Academy of Sciences, Beijing, China*

HAILING JIN • *Department of Microbiology and Plant Pathology, Center for Plant Cell Biology, Institute for Integrative Genome Biology, University of California, Riverside, CA, USA*

JAN A. L. VAN KAN • *Laboratory of Phytopathology, Wageningen University & Research, Wageningen, The Netherlands*

BOQIANG LI • *State Key Laboratory of Plant Diversity and Specialty Crops, Institute of Botany, Chinese Academy of Sciences, Beijing, China; China National Botanical Garden, Beijing, China*

KUI LIU • *State Key Laboratory of Plant Diversity and Specialty Crops, Institute of Botany, Chinese Academy of Sciences, Beijing, China; China National Botanical Garden, Beijing, China; University of Chinese Academy of Sciences, Beijing, China*

FALVIA ANGELA MARIA MAGGIOLINI • *Council for Agricultural Research and Economics–Research Center for Viticulture and Enology (CREA-VE), Via Casamassima, Turi BA, Italy*

ANTONIO DOMENICO MARSICO • *Council for Agricultural Research and Economics–Research Center for Viticulture and Enology (CREA-VE), Via Casamassima, Turi BA, Italy*

JULIÁN MÉNDEZ-GARCÍA • *Centro de Biotecnología y Genómica de Plantas, Universidad Politécnica de Madrid (UPM)-Instituto Nacional de Investigación y Tecnología Agraria y Alimentaria (INIA/CSIC), Campus de Montegancedo, Pozuelo de Alarcón, Madrid, Spain*

SIMONÉ MURGUIA • *Department of Microbiology and Plant Pathology, Center for Plant Cell Biology, Institute for Integrative Genome Biology, University of California, Riverside, CA, USA*

JAVIER PARDO-MEDINA • *Centro de Biotecnología y Genómica de Plantas, Universidad Politécnica de Madrid (UPM)-Instituto Nacional de Investigación y Tecnología Agraria y Alimentaria (INIA/CSIC), Campus de Montegancedo, Pozuelo de Alarcón, Madrid, Spain; Departamento de Genética, Facultad de Biología, Universidad de Sevilla, Sevilla, Spain*

FRANK P. J. PIETERSE • *Laboratory of Phytopathology, Wageningen University & Research, Wageningen, The Netherlands*

JULIO L. RODRÍGUEZ-ROMERO • *Centro de Biotecnología y Genómica de Plantas, Universidad Politécnica de Madrid (UPM)-Instituto Nacional de Investigación y Tecnología Agraria y Alimentaria (INIA), Campus de Montegancedo, Pozuelo de Alarcón, Madrid, Spain; Departamento de Biotecnología-Biología Vegetal, Escuela Técnica Superior de Ingeniería Agronómica, Alimentaria y de Biosistemas, Universidad Politécnica de Madrid (UPM), Madrid, Spain*

ANA RUIZ-PADILLA • *Centro de Biotecnología y Genómica de Plantas, Universidad Politécnica de Madrid (UPM)-Instituto Nacional de Investigación y Tecnología Agraria y Alimentaria (INIA), Campus de Montegancedo, Pozuelo de Alarcón, Madrid, Spain*
MAREN SCHARFENBERGER-SCHMERR • *Hochschule Kaiserslautern, Weincampus Neustadt, Neustadt, Germany*
XIAOQIAN SHI-KUNNE • *Laboratory of Phytopathology, Wageningen University & Research, Wageningen, The Netherlands*
H. M. SURAJ • *Laboratory of Phytopathology, Wageningen University & Research, Wageningen, The Netherlands*
SHIPING TIAN • *State Key Laboratory of Plant Diversity and Specialty Crops, Institute of Botany, Chinese Academy of Sciences, Beijing, China; University of Chinese Academy of Sciences, Beijing, China; China National Botanical Garden, Beijing, China*
YUAN WANG • *Department of Microbiology and Plant Pathology, Center for Plant Cell Biology, Institute for Integrative Genome Biology, University of California, Riverside, CA, USA*
JIN WEIBO • *Key Laboratory of Plant Secondary Metabolism and Regulation of Zhejiang Province, College of Life Sciences and Medicine, Zhejiang Sci-Tech University, Hangzhou, China*
HUAITONG WU • *Department of Microbiology and Plant Pathology, Center for Plant Cell Biology, Institute for Integrative Genome Biology, University of California, Riverside, CA, USA*
CHEN YA • *Key Laboratory of Plant Secondary Metabolism and Regulation of Zhejiang Province, College of Life Sciences and Medicine, Zhejiang Sci-Tech University, Hangzhou, China*
LIU YIQING • *Key Laboratory of Plant Secondary Metabolism and Regulation of Zhejiang Province, College of Life Sciences and Medicine, Zhejiang Sci-Tech University, Hangzhou, China*
MIN ZHANG • *Department of Microbiology and Plant Pathology, Center for Plant Cell Biology, Institute for Integrative Genome Biology, University of California, Riverside, CA, USA*
DIAN ZHENG • *Laboratory of Phytopathology, Wageningen University & Research, Wageningen, The Netherlands*

Check for updates

Chapter 1

Botrytis cinerea Isolation Methods

Crucitti Dalila and Pacifico Davide

Abstract

Efficient and reproducible isolation of *Botrytis* species, including *Botrytis cinerea* and *Botrytis pseudocinerea*, is essential for studies on fungal biology, epidemiology, and host–pathogen interactions. The protocols described in this chapter provide practical methods for obtaining pure cultures from symptomatic and asymptomatic plant tissues, as well as from single spores and hyphal fragments. Each approach is detailed with step-by-step instructions, highlighting critical parameters for successful isolation and culture maintenance. The symptomatic and asymptomatic tissue methods enable rapid recovery of isolates from field material, whereas the single-spore and hyphal-tip protocols ensure clonal purity and reduce contamination risks. Practical notes and troubleshooting tips are also provided to facilitate reproducibility across laboratories. These methods offer a comprehensive toolkit adaptable to different experimental needs, from basic mycological research to applied plant pathology, ensuring reliable access to genetically and phenotypically stable *Botrytis* isolates for downstream analyses.

Keywords *Botrytis cinerea*, Fungal isolation, Single-spore culture, Hyphal fragments, Gray mold

1 Introduction

Research on *Botrytis cinerea*, the causative agent of gray mold, and the closely related species *Botrytis pseudocinerea* [1], relies heavily on the availability of efficient protocols for the isolation of the pathogen from field-collected materials. An ideal isolation protocol should be rapid and simple and can be used by both trained and untrained personnel. Samples can be collected from both symptomatic and asymptomatic spore-bearing lesions (Fig. 1). Since such samples may harbor populations of isolates with different genetic and sanitary characteristics [2,3], the isolation of single spores is often necessary to obtain pure strains. Although this method is precise, it is laborious and requires advanced technical skills. A faster and less complex alternative is the isolation of hyphal fragments, which has proven to be equally efficient if applied correctly [4]. In this chapter, we will briefly report the main reliable methods for *Botrytis* spp. isolation on artificial media.

Luca Nerva and Walter Chitarra (eds.), *Botrytis cinerea*, Methods in Molecular Biology, vol. 3076,
https://doi.org/10.1007/978-1-0716-5551-1_1,

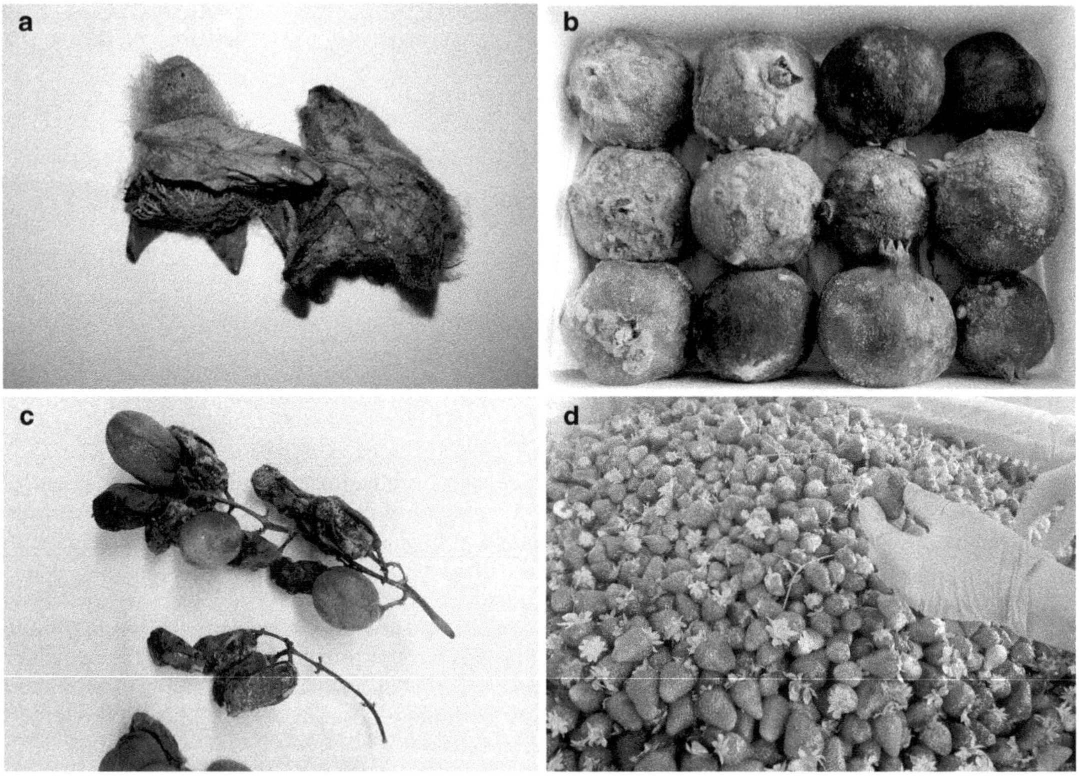

Fig. 1 *Punica granatum* flowers (**a**) and fruits (**b**), grapevine berries (**c**), and strawberry fruits (**d**) used as starting material for the *Botrytis cinerea* Symptomatic Tissue Isolation Method

2 Materials

2.1 General Lab Equipment

- Petri dishes (9 cm).
- Funnel and flasks.
- 15 mL and/or 50 mL centrifuge tubes.
- Pipettes and pipette tips.
- Inoculating loop, hockey stick, L-shaped cell spreader.
- Nylon mesh or filter.
- Counting chamber.
- Plate incubator.
- Autoclave.
- Laminar flow cabinet.
- Vortex.
- Microscope.

3 Methods

3.1 Symptomless Tissue Isolation Method

1. Disinfect plant tissues without evident symptoms by wiping with 75% ethanol for 30 s (or 1% sodium hypochlorite for 1 min) or, alternatively, by immersion in 70% ethanol for 30 s, followed by 2 min in 0.35% sodium hypochlorite and 30 s in 70% ethanol, then air-dry.
2. Place plant tissue on Petri dishes and incubate under humid conditions (97% RH, relative humidity) for 3–4 days in darkness, until the fungus grows.
3. Inoculate a small amount of mycelium of the typical *Botrytis* colony on sterilized potato dextrose agar (PDA) medium plates.
4. Incubate plates at 21°C ± 1°C for 3–7 days in darkness.
5. After fungal growth, place a mycelial disk (diameter = 4 mm) from the growing margin of the colony at the center of a new sterilized PDA plate.
6. Incubate under the same conditions as in step 4.
7. Repeat steps 5 and 6 until pure cultures are obtained.

3.2 Symptomatic Tissue Isolation Method

1. Cut diseased plant tissues with an obvious spot (gray mold) with scissors or a scalpel into 3–5 mm tissue pieces.
2. Put them directly on sterilized PDA medium plates.
3. Incubate plates at 21°C ± 1°C for 3–7 days in darkness.
4. After fungal growth, place a mycelial disk (diameter = 4 mm) from the growing margin of the colony at the center of a new sterilized PDA plate.
5. Incubate under the same conditions as in step 3.
6. Repeat steps 4 and 5 until pure cultures are obtained.

3.3 Single Spore Isolation Method

1. Inoculate stock cultures in Petri dishes on PDA medium.
2. Incubate under cool white fluorescent light (14-h photoperiod; 65 $\mu mol\, m^{-2} s^{-1}$) in a growth chamber at 18°C (night) and 23°C (day) for about 2 weeks.
3. Harvest spores by scrubbing the aerial mycelium with a pipette tip to collect a substantial amount of tissue, then suspend in 5–10 mL of sterile distilled water.
4. Filter the suspensions through nylon mesh on a funnel or through 30 μm mesh filters to remove mycelial fragments.
5. Vortex for 1 min to separate the spores.
6. Measure the concentration of spores with a counting chamber. Calibrate the final concentration of the spore suspension.

7. According to the final concentration of the stock spore solution, create a decimal serial dilution in sterile distilled water.
8. Transfer a loopful of the spore suspension from each dilution tube onto the surface of water agar plates suitable for spore germination.
9. Spread the spore suspension across the plate using a glass hockey stick or L-shaped bacteria spreader.
10. Incubate the unsealed plates at 25°C for 12–24 h.
11. Once the spores have germinated, usually after 24 h, pick up a small piece of agar containing a single germinated spore.
12. To establish that the spore is the one desired, prepare a slide and examine under the compound microscope.
13. Transfer the block of medium carrying a single spore to a fresh PDA plate and incubate at 25°C until the colony diameter is about 1–2 cm.
14. Excise a mycelial plug from the periphery of the young colony initiated by the single spore and transfer it to a fresh PDA plate.
15. Check the culture after a few days, and if there is no contamination, the purified culture can be stored.

3.4 Hyphal Fragment Isolation Method

1. Place a disc of fungal colony in the middle of a malt extract agar (MEA) plate and incubate for 1–2 days at 25°C (optimal growth conditions).
2. Place the Petri plate on the stage of a compound microscope and locate a hyphal tip using the low-power objective.
3. Excise a small block of medium carrying a fragment of hypha and transfer it to a fresh PDA plate.
4. Maintain the pure culture.

4 Recipes

4.1 Potato Dextrose Agar (PDA) from Commercial Powder

1. Add 39 g of commercial PDA powder to 1000 mL distilled water. Alternatively, add 15 g of bacteriological agar, 20 g of dextrose, and 4 g of potato starch to 1000 mL distilled water. Alternatively, PDB can be mixed with 2% bacteriological agar.
2. Heat at 80°C to dissolve the medium completely (or skip this step and go directly to step 3).
3. Autoclave for 15 min at 121°C.
4. Distribute 20–25 mL into sterile 15 × 100 mm Petri dishes or 20 mL into sterile 90 mm diameter plates.

4.2 PDA from Potato Infusion

1. Potato infusion: boil 200 g of sliced potatoes in 1 L of distilled water for 30 min.
2. Separate the infusion by filtering the entire content with sterile gauze.
3. Add water, dextrose, and agar, then boil to dissolve.
4. Autoclave for 15 min at 121°C.
5. Adjust the final to pH 5.6 ± 0.2.
6. Distribute 20–25 mL into sterile 15 × 100 mm Petri dishes.

4.3 Water Agar from Commercial Powder

1. Suspend 15 g of agar (1.5%)in 1000 mL of distilled water.
2. Heat to boiling to dissolve the medium completely.
3. Sterilize by autoclaving at 121°C for 15 min.
4. Final pH 6.3 ± 0.3 at 25°C.
5. Dispense as desired.

4.4 Malt Extract Agar (MEA) from Commercial Powder

1. Suspend 33.6 g of the medium in 1 L of distilled water. Alternatively, add 15 g of bacteriological agar, 12.75 g of maltose, 2.75 g of dextrin, 2.35 g of glycerol, and 0.78 g of peptone to 1000 mL of distilled water.
2. Mix well and dissolve by heating with frequent agitation. Boil for 1 min until complete dissolution (do not overheat).
3. Sterilize in an autoclave at 118°C for 10 min.
4. Cool to 45–50°C, mix well, and dispense into sterile 90 mm diameter plates.

5 Notes

1) It is important to pour media at no more than 60°C. Cool media at room temperature.
2) Use a sterilized scalpel or cork borer to cut mycelium plugs from fungal cultures.
3) It is recommended to sterilize filter paper at 80°C for at least 6 h.
4) For isolate collection, it is preferable to use symptomatic tissues hosting the mycelium in active proliferation visible from late spring to autumn, although the typical gray mold can also be observed on the residues of vegetation as well as on shriveling berries during the winter, particularly in warm climates.
5) When collecting and handling the plant material, wear disposable gloves to avoid sample contamination.

6) *B. cinerea* isolates can be easily propagated on PDA without the addition of acid or antibiotics.
7) The use of malt extract agar for the hyphal fragment isolation allows the growth of the fungus on this medium to be sparse, facilitating the observation of well-separated individual hyphae at the margin of the colony.
8) Water agar is recommended for enumeration, cultivation, and observation of sporulation of some fungi.
9) Although the traditional PDA preparation method based on potato infusion can also be suitable for fungal growth, it is recommended to use standard commercially available PDA media powders, preparing the mixture as indicated by the manufacturer.
10) Since PDA is not a selective medium, each plate should be frequently checked, at least every 48–72 h, to exclude contamination with other fungal or bacterial species possibly derived from the original field sample. Alternatively, a selective medium can be used.
11) Working under a hood and in aseptic conditions ensures a lower risk of contamination when transferring *B. cinerea* isolates. Use sterile glassware and single-use disposable devices or sterilize with a Bunsen burner.
12) Several re-isolation steps may be required to eliminate isolate contamination.
13) Botrytis spores can be preserved in 50% glycerol at −80°C.

References

1. Walker AS, Gautier AL, Confais J, Martinho D, Viaud M, Le Pêcheur P, Dupont J, Fournier E (2011) *Botrytis pseudocinerea*, a new cryptic species causing gray mold in French vineyards in sympatry with *Botrytis cinerea*. Phytopathology 101(12):1433–1445. https://doi.org/10.1094/PHYTO-04-11-0104
2. Leyronas C, Duffaud M, Nicot PC (2012) Compared efficiency of the isolation methods for *Botrytis cinerea*. Mycology 3(4):221–225. https://doi.org/10.1080/21501203.2012.727484
3. Zhang Y, Li X, Shen F, Xu H, Li Y, Liu D (2018) Characterization of *Botrytis cinerea* isolates from grape vineyards in China. Plant Dis 102(1):40–48. https://doi.org/10.1094/PDIS-01-17-0062-RE
4. Ruiz-Padilla A, Rodríguez-Romero J, Gómez-Cid I, Pacifico D, Ayllón MA (2021) Novel mycoviruses discovered in the mycovirome of a necrotrophic fungus. mBio 12(3):e03705–20. https://doi.org/10.1128/mBio.03705-20

Check for updates

Chapter 2

Culture and Maintenance Methods for *Botrytis cinerea*

Crucitti Dalila and Pacifico Davide

Abstract

Culturing and maintaining *Botrytis cinerea* in vitro is essential for advancing research on fungal biology, host–pathogen interactions, and plant disease management. Reliable culture methods provide the basis for reproducible experiments, ensuring stable growth and long-term preservation of isolates. This chapter details standardized procedures for establishing fungal colonies on agar and broth media, including Universal Beer Agar, Potato Dextrose Agar, and Czapek-Dox Agar, as well as specialized formulations such as Carrot Agar and Honey Peptone Agar. Protocols for short-term maintenance on agar slants and long-term storage in glycerol at −80°C are described to ensure strain viability over time. Critical variables affecting fungal growth, such as temperature, light, pH, and medium composition, are also discussed, providing guidance for optimizing culture conditions. Together, these methods serve as practical tools for laboratories aiming to maintain pure *B. cinerea* cultures for physiological, genetic, and pathological studies.

Keywords *Botrytis cinerea*, Fungal culture, Agar media, Broth cultures, Long-term storage

1 Introduction

The successful cultivation and maintenance of fungal colonies in a laboratory setting are fundamental to studying their biological properties, including growth dynamics, pathogenicity, and response to environmental variables. *Botrytis cinerea*, a pathogen notorious for its role in gray mold disease, serves as an exemplary model in mycological research due to its agricultural impact and complex life cycle. Standardized culture techniques facilitate the controlled examination of its physiology, enabling researchers to assess various growth factors, including nutrient media, temperature, pH, and light conditions. This chapter outlines protocols for the culturing and maintenance of *B. cinerea* colonies, both short-term and long-term, on different agar and broth media formulations, as well as highlights essential considerations for obtaining reliable and reproducible growth outcomes (Fig. 1).

Luca Nerva and Walter Chitarra (eds.), *Botrytis cinerea*, Methods in Molecular Biology, vol. 3076,
https://doi.org/10.1007/978-1-0716-5551-1_2,

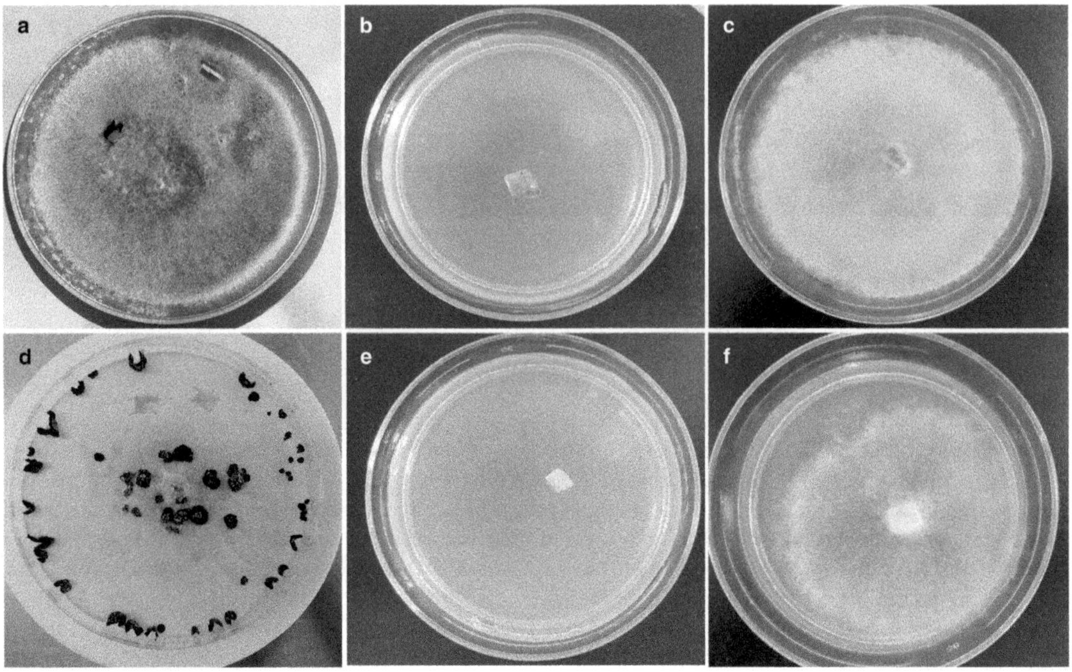

Fig. 1 Cultivation and maintenance of *Botrytis cinerea* isolates under laboratory conditions on PDA plates. Isolate CNR_IBBR BC1 at T0 (**a, b**) and after 7 days (T1; **c**). Isolate CNR_IBBR BC2 at T0 (**d, e**) and after 7 days (T1; **f**)

The selection of appropriate culture media is crucial for maximizing colony viability and development. Agar-based media, such as Universal Beer Agar (UBA), Czapek-Dox Agar (CDA), and Potato Dextrose Agar (PDA), support diverse nutrient requirements, while specialized formulations such as Honey Peptone Agar (HPA) and Carrot Agar (CA) cater to specific metabolic needs. Broth media, alternatively, provide a simpler substrate conducive to rapid conidial and hyphal growth under static incubation.

2 Materials

Universal Beer Agar (UBA) (from Commercial Powder)

1. Suspend 55 g of medium (dextrose, dipotassium phosphate, magnesium sulfate, monopotassium phosphate, sodium chloride, yeast extract, bacteriological agar, ferrous sulfate, peptonized milk, and tomato juice) in 750 mL of distilled water. Stirr and dissolve by heating with frequent agitation.
2. Boil for 1 min until complete dissolution.
3. Add 250 mL of beer without degassing when the medium is still hot. Mix with agitation.

4. Dispense into containers and sterilize in autoclave at 121°C for 10 min. Cool to 45–50°C, mix well, and dispense into plates.

Czapek-Dox Agar (CDA) (from Commercial Powder)

1. Suspend 50 g of medium (composition: sucrose, sodium nitrate, potassium chloride, magnesium sulfate heptahydrate, iron(II) sulfate heptahydrate, dipotassium hydrogen phosphate, and agar) in 1 L of distilled water.
2. Sterilize by autoclaving at 121°C for 15 min. Stir well before pouring.

Plate Count Agar (PCA) (from Commercial Powder)

1. Dissolve 22.5 g of medium (composition: tryptone, yeast extract, dextrose, and agar) in 1 L of purified water.
2. Stir until fully dissolved.
3. Autoclave at 121°C for 15 min.

Richard's Synthetic Agar (RA) (from Commercial Powder)

1. Suspend 82.52 g of medium (potassium nitrate, monopotassium dihydrogen phosphate, magnesium sulfate, ferric chloride, sucrose, and agar) in 1 L of distilled water.
2. Autoclave at 121°C for 15 min.
3. Mix well and pour into sterile Petri plates.

Honey Peptone Agar (HPA)

1. Dissolve 60 g of pure honey, 10 g of peptone, and 20 g of agar in 1 L of distilled water.
2. Stir until fully dissolved.
3. Autoclave at 121°C for 15 min.
4. Pour into sterile Petri plates.

Carrot Agar (CA) (from Commercial Powder)

1. Suspend 19 g of medium (composition: carrot infusion and agar) in 1 L of purified/distilled water.
2. Sterilize by autoclaving at 121°C for 15 min.
3. Mix well and pour into sterile Petri plates.

3 Methods

3.1 Colony Culture

<u>Using Agar Media</u>

1. In an aseptic environment, use a sterilized cork borer to cut a small cylindrical section from a Petri dish with developed mycelium.

2. Using a sterile scalpel, place the mycelium plug in the center of a prepared agar medium plate. Ensure that part of the mycelium surface contacts the culture medium.
3. Incubate the plate at 20°C under controlled temperature and lighting until radial colony growth is visible.

Using Broth Media

1. Inoculate an aliquot (200 μL) of fungal conidial suspension into a sterile broth medium (10 mL) in a 25 mL cell culture flask. Alternatively, inoculate a 1 cm plug of fungal growth culture into a 250 mL conical flask containing 250 mL of sterile broth medium.
2. Incubate cultures at 25°C ± 1°C under static conditions for a sufficient growth period (until strong turbidity of the medium is observed).
3. Separate and harvest the cultured material using a filtration method.

Briefly, fold a No. 1 Whatman filter paper in half, then fold twice more to form a 90° angle; stack three layers of filter paper on one side and open one on the other side to form a funnel. Place the paper cone into a sterile glass funnel in a sterile flask, pouring the broth culture along the inner funnel wall. Collect the filtered mycelium in the flask below.

3.2 Colony Maintenance

Short-Term Storage

1. Prepare agar slants by pouring autoclaved agar medium into culture tubes in a slanted position and allow it to solidify.
2. Once a pure colonized plate is available, transfer a section of mycelium from the center of the plate to the slant using a sterile scalpel. For broth cultures, dip a sterile swab into the fungal suspension and streak it on the slant surface.
3. Incubate under controlled conditions to encourage fungal growth.
4. Ensure that the fungus grows without contamination and store slants at 4°C.

Long-Term Storage

1. Prepare a 10%–20% glycerol solution (w/w) by measuring 50–100 mL of pure glycerol with a serological pipette and adding distilled water to a final volume of 500 mL. Use tempered borosilicate glass bottles to autoclave the glycerol solution.
2. Dispense 1 mL of sterilized glycerol solution into 2 mL cryogenic tubes sterilized by autoclaving or prolonged UV exposure.

3. Transfer small agar blocks with mycelium into cryogenic tubes and add glycerol solution until full. Alternatively, mix 1 mL of filtered fungal suspension with the 10%–20% glycerol solution to achieve a 50% mixture.
4. Acclimate the fungal isolates by placing the tubes in a refrigerator for several hours, then store them at −80°C.

4 Notes

- **Effect of Media:** *B. cinerea* colony color changes from white to gray over time, with radial growth rates varying across media. Higher mycelial and sclerotial growth rates were observed on UBA and PDA media compared to CDA and PCA in dark conditions [1]. The highest mycelial growth was on RA medium, followed by HPA and PDA, with the lowest on CA medium [2].
- **Effect of Incubation Temperature:** Black sclerotial formation is optimal in darkness at 20°C for 6–7 days. Sclerotial formation is slower at 10°C regardless of light or agar medium. Radial mycelial growth increases from 5°C to 20°C, ceasing at 35°C. The largest colony diameter is observed at 20°C, decreasing as temperature rises.
- **Effect of Light on Growth:** *B. cinerea* is light-sensitive, with different wavelengths affecting spore germination, colony expansion, and mycelial accumulation [3]. Red, blue, and white light, as well as darkness, have similar sporulation and growth effects, while green light reduces growth. Under green light, mycelial growth is slow, while under near UV or darkness, it spreads completely over the medium.
- **Effect of pH Medium:** Both mycelial growth and sclerotial germination of *B. cinerea* are influenced by the pH of the culture medium. Fungus growth is extensive on PDA at pH 4.5, covering plates entirely at pH 6.0 after incubation at 25°C. Alkaline pH (7.4–8.2) significantly reduces or halts mycelial growth, and no sclerotial germination is observed at pH 8.0 and above after 6 days of incubation; no growth is recorded.

References

1. Rasiukevièiûtë N, Moroèko-Bièevska I, Audrius Sasnauskas A (2017) Characterisation Of growth variability and mycelial compatibility of *Botrytis cinerea* isolates originated from apple and strawberry in Lithuania. Proc Latv AcadSci 71(3):217–224. https://doi.org/10.1515/prolas-2017-0036
2. Sultana S, Sikder MM, Ahmmed MS, Sultana A, Alam N (2020) Mycelial growth and biological control measures of *Botrytis cinerea* isolated from strawberry fruit rot disease in Bangladesh. Univ J Microbiol Res 8:13–18
3. Schumacher J (2017) How light affects the life of Botrytis. Fungal Genet Biol. 106:26–41

Chapter 3

In Vitro Sexual Reproduction of *Botrytis cinerea*

Dian Zheng and Jan A. L. van Kan

Abstract

Sexual reproduction is a crucial aspect of the lifecycle of *Botrytis cinerea*. This protocol outlines a method to induce sexual reproduction of *B. cinerea* under laboratory conditions and obtain ascospores from which sexual offspring can be grown. A video was prepared that illustrates the entire process.

Keywords Sexual reproduction, *Botrytis cinerea*, Sclerotia, Apothecium, Ascospores

1 Introduction

Botrytis cinerea is a heterothallic fungus, meaning that isolates with one mating type (MAT1-1) require a partner with the opposite mating type (MAT1-2) to initiate sexual reproduction. The MAT loci of *B. cinerea* are located at the same position on chromosome 1 but lack sequence similarity and are referred to as "idiomorphs" [1]. The MAT1-1 idiomorph encodes two genes: *MAT1-1-1*, which contains a conserved MAT-α domain, and *MAT1-1-5*, which has no known protein domain. The MAT1-2 idiomorph also contains two genes: *MAT1-2-1*, which includes a conserved HMG box, and *MAT1-2-10*, which lacks a known domain [2]. As in most Ascomycota, the MAT locus of *B. cinerea* is flanked by the genes *APN2* (encoding a DNA lyase) and *SLA2* (encoding a cytoskeletal protein), which are transcribed convergently towards the MAT locus [2, 3]. The mating type of a strain can be identified by PCR using idiomorph-specific primer sets [3] or by examining genome data. Strains with the *MAT1-2-1* and *MAT1-2-10* genes are MAT1-2, while those with the *MAT1-1-1* and *MAT1-1-5* genes are MAT1-1. Commonly used *B. cinerea* isolates, such as B05.10 and SAS56, are MAT1-1 strains, whereas SAS405 is a MAT1-2 strain [4].

A study on transcriptomes of sclerotia, three stages of apothecium development (primordia, stipes, and apothecial disks), and ascospores has been published, and the data has been deposited in the NCBI

Luca Nerva and Walter Chitarra (eds.), *Botrytis cinerea*, Methods in Molecular Biology, vol. 3076,
https://doi.org/10.1007/978-1-0716-5551-1_3,

SRA archive under PRJNA351919 (SRP093589) [4]. Numerous genes encoding virulence factors are upregulated in ascospores as compared to mature apothecial disks, which reveals that ascospores are transcriptionally primed for infection prior to their arrival on a host plant [4]. Besides, at the initiation and completion of the sexual cycle, transcriptional changes were reported to occur in physically clustered genes, suggesting regulation by epigenetic changes as discussed in detail in [4]. A population genomic study revealed a strong decay of linkage disequilibrium, suggesting extensive sexual reproduction inducing recombination amongst natural strains despite the paucity of reports of *Botrytis cinerea* apothecia in the wild [5,6]. Similarly, in *Botrytis elliptica*, linkage disequilibrium analyses indicate that between 9.4% and 19.3% of the loci in clone-corrected samples are linked, which suggests sexual recombination also occurs in the *B. elliptica* population [7]. However, in *Botrytis tulipae*, linkage disequilibrium analyses indicated that between 25.2% and 48.6% of the loci in clone-corrected samples were linked, which suggests that the *B. tulipae* population is mainly clonal with a low rate of sexual recombination [7].

The protocol below is derived from the protocol by Faretra and co-workers [8], who were pioneers in developing sexual crosses in the lab. The protocol has been used to analyze the sexual fertility of a number of mutants in fungal genes [9,10]. Besides, this protocol can also be applied in other *Botrytis* species, such as *Botrytis elliptica* [11].

2 Materials and Equipment

2.1 Materials

- Sterile water.
- Toothbrush (soft).
- Tweezers.
- Sterile Petri dishes.
- Six-well microtiter plates.
- Malt Extract Agar (MEA) plates.
- Scalpel.
- Sterile glass wool.
- 50 mL tubes.

2.2 Equipment

- 20°C dark incubator.
- 0°C dark incubator.
- 13°C incubator with 12-h light/dark cycles.

3 Methods

3.1 Generation of Microspores and Sclerotia

The generation of microspores and sclerotia requires nutrient starvation, darkness, and cold treatment.

1. Inoculate parental strains on Malt Extract Agar (MEA) plates. Place the plates in a box, and wrap the box with aluminum foil to prevent light exposure, while still ensuring proper aeration of the culture. Incubate at 20°C for 4 weeks.
2. After 4 weeks, sclerotia will form, appearing as black aggregated structures, sometimes with shiny water droplets on their surfaces (Fig. 1). These sclerotia are not fully mature and require cold treatment before being used in a cross. Transfer the plates to a 0°C incubator for 4 weeks to mimic winter conditions. After this period, the sclerotia and microspores will be ready for use.

3.2 Harvesting Sclerotia

1. In a laminar flow cabinet, use tweezers to gently remove the sclerotia from the agar. Handle carefully to minimize additional damage to the agar surface.
2. Collect the sclerotia from one culture in a Petri dish with 10–20 mL sterile water.
3. Clean the sclerotia by holding them with tweezers and brushing their surfaces with a sterile soft toothbrush in the sterile water. This removes small residues of agar and the water-repellent layer (containing hydrophobin proteins) on the aerial surface of the sclerotia that can hinder interaction between sclerotia and microspores. Proper cleaning makes the sclerotia more wettable.

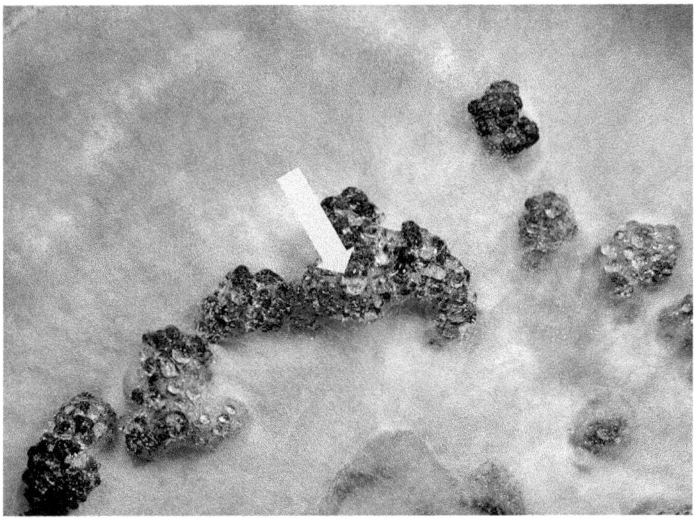

Fig. 1 Sclerotia of *Botrytis cinerea*. Water droplets (pointed out by white arrow) are formed on the hydrophobic surface of sclerotia

4. After cleaning all sclerotia from one plate, transfer them with tweezers to a different Petri dish with clean sterile water. Discard the first Petri dish with the dirty rinsing fluid.

3.3 Harvesting Microspores

1. In a laminar flow cabinet, add 5–10 mL of sterile water to the plates from which sclerotia have been removed. Use a (plastic or glass) spreader to gently scrape the surface and release the microspores. Avoid breaking off chunks of agar from areas where sclerotia were collected.
2. Transfer the suspension of mycelial fragments and microspores to a 50 mL tube and dilute with sterile water to a final volume of 15 mL.
3. Repeat steps 3.2 and 3.3 to harvest sclerotia and microspores from the partner strain (of the opposite mating type). Use sclerotia from one strain and microspores from the partner strain to set up the cross.

3.4 Setting Up the Cross

1. In a laminar flow cabinet, use tweezers to transfer sclerotia into a six-well plate, placing between one and five sclerotia in each well, depending on size and available number.
2. Using a sterile pipette, mix the suspension of mycelial fragments and microspores by pipetting up and down several times before transferring 3 mL aliquots to five of six wells in the plate. Treat the sclerotium in the sixth well with 3 mL sterile water as a negative, nonfertilized control.
3. Seal the six-well plate with parafilm and incubate at 13°C with dim white light for 12 h per day from the top. Every week, gently move the plates to agitate any microspores that may have developed but not yet been in contact with the sclerotial surface. Stipes will emerge from the bottom edge of the sclerotia after 30–40 days and grow toward the light source (Fig. 2A). Apothecia should be observable after 60–90 days (Fig. 2B). Multiple apothecia can develop on a single sclerotium.

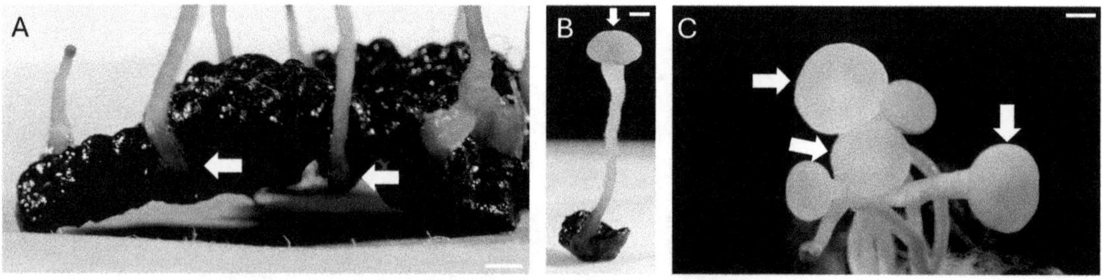

Fig. 2 Apothecium of *Botrytis cinerea*. (**A**) Stipes (pointed by white arrow) emerging from sclerotia at early stage of sexual reproduction. (**B**) Fully developed apothecium (pointed by white arrow) at late stage of sexual reproduction. (**C**) Top view of a mature apothecial disks. White bar, 1 mm

3.5 Sampling Ascospores

1. Select a mature apothecium and place it on a sterile empty Petri dish.
2. Use a scalpel to cut a disk from the apothecium (Fig. 2C) and transfer it to a sterile mortar.
3. Add 5 mL of sterile water and gently crush the apothecial disk with a sterile pestle to release the asci and ascospores. At this stage, samples can be taken for examination by microscopy.
4. When the entire apothecial disk appears to be fragmented and forms a seemingly homogeneous (yet turbid) suspension in water, filter the ascospore suspension through a small piece of sterile glass wool in a 5 mL pipette tip into a 50 mL tube to separate the ascospores from the large debris of apothecial disk tissue.
5. Add sterile water to a final volume of 30–50 mL and centrifuge at 250 × g for 10 min to pellet the ascospores. A white pellet should be visible after centrifugation; the size of the pellet depends on the number of apothecia that are sampled. One single apothecium only forms a tiny pellet.
6. Remove the supernatant and resuspend the pellet in an appropriate volume of sterile water based on experimental requirements. The ascospores are now ready for use or storage in 20% glycerol (see Pieterse and van Kan, Chapter 4 in this book).

A video demonstrating the entire process can be found on the following website:

https://www.wur.nl/en/Research-Results/Chair-groups/Plant-Sciences/Laboratory-of-Phytopathology.htm

4 Notes

1. The strain from which the sclerotia will be used in the cross is referred to as the maternal parent. The strain from which the microspores will be used is the paternal parent. Both types of material can be generated from the same strain on the same plate; however, strains of the same mating type cannot perform sexual reproduction.
2. Do not place all strains for crossing in the same box. It is better to keep cultures of mating partners in separate boxes to prevent contamination, especially during formation of microspores.
3. After using the toothbrushes, soak them in 70% ethanol for a few nights and then dry them in a 60°C oven to sterilize them before the next use. Otherwise, microspores might survive in the bristles, leading to contamination in subsequent

experiments. For one experiment, it is best to have a set of toothbrushes.

4. The workbench should be thoroughly cleaned and sterilized between separate crosses to remove all microspores that may have landed somewhere on the tools in the flow cabinet and may contaminate the next cross.
5. Avoid contact between the plate and your sleeves, as microspores on sleeves can result in contamination when setting up other crosses.
6. The development of fruiting bodies is sensitive to temperature. If the temperature rises above 15°C, the sclerotia may start to produce mycelium and conidiospores instead of apothecia.
7. When selecting sclerotia from the plate, prioritize large sclerotia. Some small sclerotia may not be able to develop mature apothecia due to a lack of nutrient resources.
8. Light in the 13°C incubator is crucial for the development of apothecia. Do not use LED light, since we have observed a failure to produce mature apothecia under LED light with a natural sunlight spectrum. It is unclear whether this was due to the spectrum or the intensity. A classical white light tube (Philips MASTER TL-D Super 80 18W-830 Warm White | 60 cm) is suitable for apothecium formation. The distance between the light tube and the six-well microtiter plates should be around 25 cm to prevent condensation on the plate lids caused by heat from the light tube when positioned too closely.

Acknowledgments

We dedicate this publication to prof. Franco Faretra (University of Bari Aldo Moro, Italy) who passed away in August 2026. Prof. Faretra was a pioneer in genetic studies on *Botrytis cinerea* and he developed the first version of the protocol that we describe in this chapter. We appreciate the help from Laurens Deurhof in generating and editing the video for this protocol.

References

1. Metzenberg RL, Glass NL (1990) Mating type and mating strategies in *Neurospora*. Bioessays 12(2):53–59
2. Amselem J, Cuomo CA, van Kan JA, Viaud M, Benito EP, Couloux A et al (2011) Genomic analysis of the necrotrophic fungal pathogens *Sclerotinia sclerotiorum* and *Botrytis cinerea*. PLoS Genet 7(8):e1002230
3. De Miccolis Angelini RM, Rotolo C, Pollastro S, Faretra F (2016) Molecular analysis of the mating type (MAT1) locus in strains of the heterothallic ascomycete *Botrytis cinerea*. Plant Pathol 65(8):1321–1332
4. Rodenburg SYA, Terhem RB, Veloso J, Stassen JHM, van Kan JAL (2018) Functional analysis of mating type genes and transcriptome analysis during fruiting body development of *Botrytis cinerea*. mBio 9(1):e01939-17
5. Atwell S, Corwin JA, Soltis N, Zhang W, Copeland D, Feusier J et al (2018) Resequencing

and association mapping of the generalist pathogen *Botrytis cinerea*. https://www.biorxiv.org/content/10.1101/489799v1
6. Mercier A, Simon A, Lapalu N, Giraud T, Bardin M, Walker AS et al (2021) Population genomics reveals molecular determinants of pecialization to tomato in the polyphagous fungal pathogen *Botrytis cinerea* in France. Phytopathology 111(12): 2355–2366
7. Staats M, van Baarlen P, van Kan JAL (2007) AFLP analysis of genetic diversity in populations from the Netherlands. Eur J Plant Pathol 117(3):219–235
8. Faretra F, Antonacci E, Pollastro S (1988) Sexual behaviour and mating system of *Botryotinia fuckeliana*, teleomorph of *Botrytis cinerea*. J Gen Microbiol 134:2543–2550
9. Segmuller N, Kokkelink L, Giesbert S, Odinius D, van Kan J, Tudzynski P (2008) NADPH oxidases are involved in differentiation and pathogenicity in *Botrytis cinerea*. Mol Plant Microbe Interact 21(6):808–819
10. Jonkers W, Van Kan JAL, Tijm P, Lee YW, Tudzynski P, Rep M et al (2011) The FRP1 F-box gene has different functions in sexuality, pathogenicity and metabolism in three fungal pathogens. Mol Plant Pathol 12(6):548–563
11. Terhem RB, Staats M, van Kan JAL (2015) Mating type and sexual fruiting body of *Botrytis elliptica*, the causal agent of fire blight in lily. Eur J Plant Pathol 142(3):615–624

Check for updates

Chapter 4

Viable Storage of *Botrytis* Conidia and Sclerotia

Frank P. J. Pieterse and Jan A. L. van Kan

Abstract

Long-term storage of microorganisms is essential for maintaining a consistent laboratory strain collection. To prevent a reduction in viability of *Botrytis* stocks, we can make use of the naturally formed survival structures of this fungus, conidiospores and sclerotia. Conidiospore stocks are the preferred method of storage and can be stored in 20% glycerol at −80°C for up to 30 years. When conidiospore storage is not feasible, sclerotia can be stored at 0°C, and agar plugs containing mycelium can be stored at 4°C. Sclerotia and mycelium stocks will only last several years and thus require more frequent propagation.

Keywords Long-term storage, Glycerol stock, Conidia, Sclerotia

1 Introduction

When continuously propagating cultures of microbes, mutations will occur, gradually changing the genotype of laboratory strains [1]. To prevent this issue, it is important to maintain proper long-term stocks, which allow researchers to revive and study the same genotype continuously. It is essential to work carefully and properly when handling long-term storage, as the loss of stocks can be disastrous to research on microbes.

Most long-term storage protocols rely on storing cells in a glycerol solution in −80°C freezers. For *Botrytis*, long-term stocks can be made from conidiospores in 20% (v/v) glycerol/water [2]. The absence of nutrients in these stocks prevents premature germination of the spores [3]. Therefore, viability of these stocks remains relatively constant over time, and stocks can remain viable for up to 30 years. Each time a glycerol stock is thawed, its viability may be reduced, and a risk of contamination is introduced [1]. This is especially problematic for strains that will be revived numerous times, such as standard wild-type laboratory strains.

For situations where −80°C freezers are not available, or when storing mutants deficient in conidiospore formation, alternative storage methods can be used that rely on sclerotia or mycelial plugs.

Luca Nerva and Walter Chitarra (eds.), *Botrytis cinerea*, Methods in Molecular Biology, vol. 3076,
https://doi.org/10.1007/978-1-0716-5551-1_4,

Sclerotia are macroscopic survival structures that *Botrytis* forms to survive the winter in the absence of living plant hosts [4]. In the development of sclerotia, water is exuded from the structure [5,6], which may allow them to survive in <0°C temperatures without freezing [7]. Sclerotia can be easily produced even outside a laboratory setting by simply growing *Botrytis* for several weeks in complete darkness. Sclerotium development may be more pronounced when cultures are grown at lower temperatures (12°C or 15°C). Entire plates containing sclerotia can be stored at 0°C for at least 3 years. However, sclerotia-containing plates are not devoid of nutrients and can thus be contaminated by other saprophytic microbes, such as *Penicillium* sp. To prevent contamination, sclerotia can be harvested using tweezers and stored separately. As only a few dozen sclerotia are formed on each plate, loss of viability will much more quickly lead to loss of the stock than when using conidia. Therefore, it is important to transfer the sclerotia to fresh plates every year to maintain the stock. In situations where conidial glycerol stocks and sclerotia stocks cannot be made, for instance when using a strain deficient in the formation of either structure, mycelial plugs can be stored. Mycelial plugs are stored in water to prevent drying out. Mycelium can be stored in water for several years, but viability could be reduced over longer periods [8]. Therefore, it is important to renew these stocks every year.

2 Materials

2.1 Glycerol Stock of Conidiospores

To prevent premature germination and contamination, use demineralized water for producing glycerol stock solutions.

1. Glycerol stock solution (60% vol/vol glycerol and demineralized water).
2. Conidiospore suspension in nutrient-free demineralized water (>1000 spores/μL).
3. Screw-cap 1.5 mL tubes.

2.2 Storage of Sclerotia

1. Malt Extract Agar (MEA) plate.
2. Parafilm.
3. Sterile tweezers.
4. Screw-cap 1.5 mL tubes.

2.3 Storage of Mycelial Plugs

1. Malt Extract Agar (MEA) plate.
2. Square-end spatula or scalpel or agar punch.
3. Sterile 15 or 50 mL tubes.
4. Sterile demineralized water.

3 Methods

3.1 Storage of Conidiospore Glycerol Stocks

To produce a glycerol stock of conidiospores, conidiospores must first be harvested from sporulating plates according to the protocol outlined in Chap. 3 of this book. When harvesting spores for the purpose of producing glycerol stocks, it is important to produce a high concentration of conidiospores (>1000 spores/μL) and to perform the conidiospore wash step twice. When stored in this way, conidiospores can retain viability for up to 30 years. Stocks can be thawed and refrozen multiple times.

3.2 Making Glycerol Stocks

1. Add 50 μL of 60% vol/vol glycerol stock solution to the 1.5 mL screw-cap tube.
2. To the glycerol, add 100 μL twice-washed spore solution (>1000 spores/mL).
3. Mix the stock solution by vortexing the tubes for several seconds.
4. Optional: when frequent use of the stock is expected, divide the produced glycerol stock into 10 μL aliquots in PCR tubes. These PCR tubes are for single use and can be discarded after thawing and sampling.
5. Store the tube at −80°C.

3.3 Thawing Glycerol Stocks

1. Thaw the glycerol stock at room temperature.
2. Vortex the tube to homogenize the mixture.
3. Take 0.5–1 μL and transfer to a fresh agar plate.

3.4 Storage of Sclerotia

1. Inoculate a fresh MEA plate with *Botrytis* spores and allow the mycelium to grow at 20°C in complete darkness until black sclerotia have visibly formed on the surface. For *B. cinerea* B05.10, this takes ~4 weeks. Sclerotium development may be more pronounced when cultures are grown at lower temperatures (12°C or 15°C), but it may take longer to develop mature sclerotia. The procedure may require optimization when working with new strains.
2. Tightly seal the plates with parafilm.
3. Move the plates to 0–4°C for another 4 weeks in darkness to allow the sclerotia to mature.
4. Optional: Use sterilized tweezers to remove sclerotia from the plate and transfer to sterile Eppendorf tubes.
5. Store sclerotia at 0–4°C.

3.5 Storage of Mycelial Plugs

1. Inoculate a fresh MEA plate with *Botrytis* spores and allow the mycelium to grow in darkness until it reaches the edge of the plate.
2. Using a square-end spatula, scalpel, or agar punch, cut small (~1 cm^2) sections from the agar.
3. Scoop the agar plugs into a sterile 15 or 50 mL tube.
4. Fill the tubes with sterile demineralized water until all agar plugs are submerged.
5. Store the tube at 4°C.

4 Notes

1. It is recommended to label all long-term microorganism stocks using labels printed with label printers. Printed labels are easier to read and standardize, which is especially important when multiple researchers make use of these stocks [9].
2. When producing glycerol stocks, a 60% vol/vol glycerol stock solution is used, rather than pipetting 100% glycerol directly. Due to the high viscosity, pipetting errors will inevitably occur when attempting to pipette 100% glycerol, which results in inconsistent glycerol concentrations, affecting stock viability.
3. It is important that produced glycerol stocks are frozen slowly, rather than using liquid nitrogen to rapidly freeze the stock. Rapid freezing can cause rupturing of the conidiospores, which would instantly kill them. Thawing, on the other hand, can be done more rapidly without reducing viability [10].
4. When reviving glycerol stocks, it is important to properly mix the stock prior to transferring to a fresh agar plate. Over time, conidiospores may sink to the bottom of the stock solution, forming a pellet. A sample taken from the top of the stock solution will then not contain any spores. Moreover, taking a sample from the bottom of the stock by sticking the pipette tip to the bottom of the solution risks contaminating the stock.
5. Screw-cap tubes are commonly used for storing glycerol stocks, as opposed to hinge-lid tubes, as these tubes have a lower risk of opening unexpectedly or when accidentally dropped on the floor, which would result in the contamination or loss of the sample.
6. In case of power failure, −80°C freezers may shut off, causing the deterioration of glycerol stocks, depending on the duration of the problem. To prevent loss of samples, it is recommended to maintain additional backup stocks stored at a different location. Unless a second −80°C freezer facility is readily available,

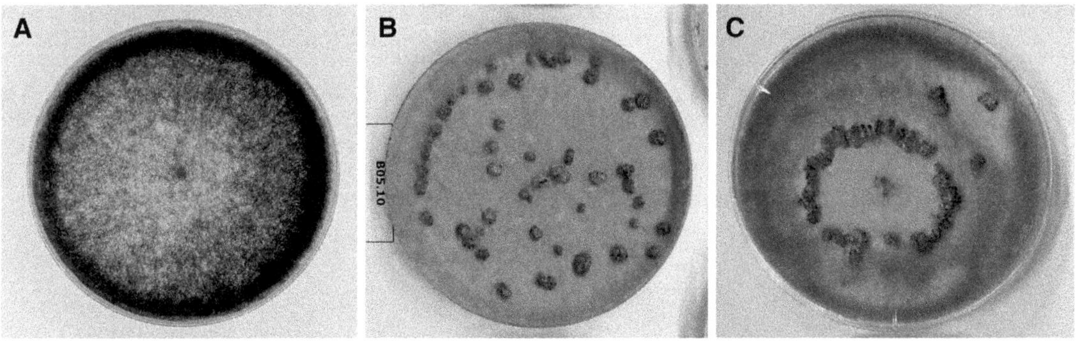

Fig. 1 *Botrytis* materials used for long-term storage. (**A**) Dark-gray conidia on a sporulating agar plate. (**B**) Black sclerotia on an agar plate inoculated with a mycelial plug. (**C**) Ring of black sclerotia on an agar plate inoculated with a glycerol stock of *Botrytis* conidia

stocks of sclerotia stored at 0°C or glycerol stocks stored at −20°C can also be used as backups.

7. Stored mycelial agar plugs can be more easily contaminated during storage. Additionally, plugs can lose viability more quickly than sclerotia or conidia. Therefore, always store multiple tubes of the same strain in parallel when using this storage method.
8. When reviving a glycerol stock for the purpose of producing sclerotia, *Botrytis* may produce a ring of sclerotia (Fig. 1C). To obtain nicely dispersed sclerotia, transfer an agar plug of growing mycelium to a fresh plate (Fig. 1B).

Acknowledgments

The research of F.P.J. Pieterse is supported by NWO grant OCENW.M.22.213.

References

1. Prakash O, Nimonkar Y, Shouche YS (2013) Practice and prospects of microbial preservation. FEMS Microbiol Lett 339(1):1–9
2. Delcan J, Moyano C, Raposo R, Melgarejo P (2002) Storage of Botrytis cinerea using different methods. J Plant Pathol 84(1):3–9.
3. Blakeman JP (1975) Germination of *Botrytis cinerea* conidia in vitro in relation to nutrient conditions on leaf surfaces. Trans Br Mycol Soc 65(2):239–247
4. Coley-Smith JR, Cooke RC (1971) Survival and germination of fungal sclerotia. Annu Rev Phytopathol 9(1):65–92
5. Sommer NF, Fortlage RJ, Buckley PM, Maxie EC (1972) Comparative sensitivity to gamma radiation of conidia, mycelia, and sclerotia of *Botrytis cinerea*. Radiat Bot 12(2):99–103
6. Willetts HJ, Bullock S (1982) Studies on the ontogeny and ultrastructure of the sclerotium of *Botrytis cinerea* Pers. ex Nocca & Balbis. Can J Microbiol 28(12):1347–1354
7. Magan N (2007) Fungi in extreme environments. In: Kubicek C, Druzhinina I (Eds.) Environmental and Microbial Relationships. The Mycota. Springer, Berlin, vol 485–103

8. Borman AM, Szekely A, Campbell CK, Johnson EM (2006) Evaluation of the viability of pathogenic filamentous fungi after prolonged storage in sterile water and review of recent published studies on storage methods. Mycopathologia 161(6):361–368
9. Berezin CT, Aguilera LU, Billerbeck S, Bourne PE, Densmore D, Freemont P et al (2023) Ten simple rules for managing laboratory information. PLoS Comput Biol 19(12): e1011652
10. Smith D, Ryan J, Stackebrandt E (2008) The ex situ conservation of microorganisms: aiming at a certified quality managementDoelle HW, DaSilva EJ (Eds.) Biotechnology. EOLSS Publisher, Oxford

Check for updates

Chapter 5

Fluorescence and Ultrastructural Microscopy for *Botrytis cinerea*

Kui Liu, Xinhua Huang, Yong Chen, Boqiang Li, Shiping Tian and Tong Chen

Abstract

Botrytis cinerea, a notorious pathogenic fungus, leads to postharvest decay in horticultural crops and substantial economic losses, while the pathogenicity of *B. cinerea* is closely related to its structural characteristics and cellular homeostasis. Among the currently available techniques, fluorescence and ultrastructural observations serve as crucial tools for investigating subcellular structures and cellular events during *B. cinerea* growth. These techniques enable the monitoring of cell structural integrity, such as variations in the cell membrane and cell wall, as well as intracellular changes, such as reactive oxygen species bursts and cell apoptosis. The results at cytological levels may not only facilitate close-up dissection of *B. cinerea* cells, but also provide important cues to uncover the inhibitory mechanism of exogenous treatments. This chapter introduces fluorescence staining and observation, as well as the preparation of ultrathin sections of *B. cinerea* conidia, aiming to provide methodological guidance for cytological examinations of *B. cinerea*.

Keywords *Botrytis cinerea*, Fluorescence observation, Ultrastructural microscopy, Structural integrity, Cellular homeostasis

1 Introduction

Botrytis cinerea, a typical necrotrophic pathogen, is capable of infecting seedlings, flowers, fruit and storage organs, resulting in postharvest losses to over 1400 species [1,2]. During the past several decades, researchers have endeavored in studies on the pathogenesis of *B. cinerea* and thus made rapid progress [2–4]. However, there is still a long way to go to get familiar with the genuine details in the life cycle of *B. cinerea*. This should be partly attributed to the lack of state-of-the-art techniques and standard experimental protocols. Moreover, our understanding of the structural and functional characteristics of fungal pathogens is still far from complet.

As an excellent example for analyzing gene functions in *B. cinerea*, the deletion of *BcSfb3*—a gene encoding a subunit of COPII vesicles—affected vegetative growth, conidiation, conidial morphology,

Luca Nerva and Walter Chitarra (eds.), *Botrytis cinerea*, Methods in Molecular Biology, vol. 3076,
https://doi.org/10.1007/978-1-0716-5551-1_5,

and reduced pathogenicity of *B. cinerea* in tomato and mung bean leaves. Furthermore, fluorescence staining and transmission electron microscopy (TEM) revealed that the *BcSfb3* deletion promoted ER-phagy and simultaneously impaired cell membrane integrity [5]. The results suggested that the full virulence of *B. cinerea* was closely related to its cellular homeostasis and structural integrity. When *B. cinerea* encounters unfavorable conditions, its pathogenicity may be diminished owing to the disruption of cellular homeostasis and damage to cellular structure [6]. Moreover, fluorescence microscopy and TEM are also important tools for checking the efficacy of exogenously applied substances on *B. cinerea*. Honokiol, a polyphenolic compound derived from *Magnolia officinalis*, markedly inhibited mycelial growth and abolished the virulence of *B. cinerea* on harvested fruit by triggering intracellular reactive oxygen species (ROS) accumulation and apoptosis. Furthermore, as revealed by fluorescence labeling and detection, honokiol effectively nullified the mitochondrial membrane potential [7]. These findings suggested that alterations in cellular homeostasis of *B. cinerea* can be readily detected by specific fluorescent dyes.

Alternatively, it was found from TEM observations that *B. cinerea* conidia from the control group exhibited normal cellular structures. The boundaries between the plasma membrane and cell wall were clearly defined, and the mitochondria displayed intact structures with well-preserved cristae. In contrast, when *B. cinerea* conidia were exposed to *Sapindus mukorossi* saponin, the boundary between the plasma membrane and cell wall became blurred. While mitochondria with intact structures and oil droplets were rarely observed, they were frequently associated with autophagic-like vacuoles (Fig. 1A). Fluorescence staining can indirectly reflect alterations in cell vitality and intracellular structures. In the *S. mukorossi* saponin group, a higher percentage of *B. cinerea* conidia were stained red by PI (propidium iodide), while the cells stained green by FDA (fluorescein diacetate) decreased (Fig. 1B, C). These findings suggest that *S. mukorossi* saponin damaged the conidia membrane and led to cell death, aligning with observations via TEM. In the treatment group, autophagic vesicles, as observed under TEM, were also stained green by MDC (monodansylcadaverine) (Fig. 1D). Consequently, *S. mukorossi* saponin disrupted the cellular structure of *B. cinerea*, thereby diminishing its pathogenicity on strawberry fruit [8]. This evidence underscores the efficacy of TEM in conjunction with fluorescence staining for accurately reflecting changes in the internal structure of *B. cinerea* under stress conditions.

In conclusion, fluorescence and ultrastructural microscopy provide robust evidence for assessing structural integrity and cellular homeostasis of *B. cinerea*. This chapter attempts to contribute experimental protocols for fluorescence detection and preparation of ultrathin sections of *B. cinerea*, aiming to provide references for advancing cytological studies on *B. cinerea*.

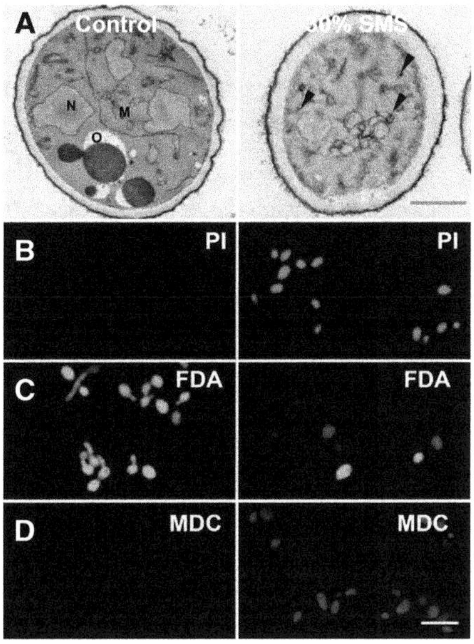

Fig. 1 Detection of the changes in cellular structure of *B. cinerea* conidia following treatment with *Sapindus mukorossi saponin*. (**A**) Ultrathin sections of *B. cinerea* as captured by transmission electron microscopy. The control conidia exhibited a dense cytosol, interspersed with numerous organelles. Conidia incubated with *S. mukorossi* saponins (SMS) displayed a significantly lighter-stained cytosol with fewer organelles. These organelles had lost their structural integrity, making them difficult to discern. In comparison to the normal structures of mitochondria in the control conidia, those in the treated conidia demonstrated an autophagic-like fusion with other vacuoles, as denoted by arrows (Bar = 2 μm). M, mitochondria; O, oil body; N, nucleus. (**B**) PI staining indicated that the membrane integrity of *B. cinerea* conidia was compromised by *S. mukorossi* saponins, with only a few conidia remaining unstained. (**C**) The FDA staining revealed that, in the control group, almost all conidia were stained by FDA, whereas only a few conidia were stained following treatment with *S. mukorossi saponins*. (**D**) *S. mukorossi saponins* notably enhanced the fluorescence intensity of MDC in *B. cinerea* cytosol (Bar = 20 μm) [8]

2 Materials

2.1 Solutions

1. Potato dextrose agar (PDA) or potato dextrose broth (PDB): Thoroughly clean and peel the potatoes, weigh about 200 g, and cut them into small pieces of 1 cm^3. Boil with 1 L distilled water for 30 min, then filter through an eight-layered gauze. Collect the filtrate and add 20 g glucose and 15 g agar powder to obtain PDA, or alternatively, add 20 g glucose to obtain PDB. Metered to a final volume of 1 L with distilled water. Autoclaved at 121 °C, 0.1 MPa for 20 min.
2. 0.1 M phosphate buffer saline (PBS, pH 7.0): Dissolve 3.561 g $Na_2HPO_4{\cdot}2H_2O$ in distilled water to a final volume of 100 mL

as solution A. Dissolve 2.76 g $NaH_2PO_4 \cdot H_2O$ in distilled water to a final volume of 100 mL as solution B. Mix 61 mL solution A and 39 mL solution B, then add distilled water to 200 mL to achieve 0.1 M phosphate-buffered saline (pH 7.0).

3. Fluorescent dyes: Prepare the solutions of fluorescent dyes according to Table 1 (*see* **Note 1**).
4. 0.2 M sodium cacodylate buffer (pH 7.2): Dissolve 4.28 g sodium cacodylate in distilled water to a final volume of 100 mL, resulting in solution A. Prepare solution B by mixing 10 mL HCl (36%–38%) with 593 mL distilled water. Combine 50 mL solution A with 4.2 mL solution B, and adjust the final volume to 100 mL with distilled water to prepare sodium cacodylate buffer (0.2 M, pH 7.2).
5. Fixation solution 1: Freshly prepare 25% glutaraldehyde and 16% paraformaldehyde before use. Combine 2.5 mL 0.2 M sodium cacodylate buffer with 0.6 mL 25% glutaraldehyde, then add 0.94 mL 16% paraformaldehyde. Finally, add 0.96 mL sterile water to create the fixation solution 1 (*see* **Note 2**).
6. Fixation solution 2: Dissolve 1.5 g $KMnO_4$ in 100 mL sodium cacodylate (0.1 M, pH 7.2) (*see* **Note 3**).
7. Spurr low-viscosity embedding agent: Combine 1 g vinylcyclohexane dioxide (VCD), 0.7 g diglycidyl ether of polypropylene glycol (DER736), 2.6 g nonenyl succinic anhydride (NSA), and 0.4 g dimethylaminoethanol (DMAE).

Table 1
List of fluorescence dyes used in this protocol

Fluorescence dyes	Stock concentration	Working concentration	Solvent	Observation condition
FDA	5 mg/mL	20 mg/L	Acetone	Channel 2EX BP 450-490FT 510, LT 515
PI	5 mg/mL	20 mg/L	Water	Channel 3 EX BP 546/12 FT 580, LP 590
DCHF-DA	10 mM	10 μM	DMSO	Channel 2 EX BP 450-490 FT 510, LT 515
MitoTracker Red	1 mM	1 μM	DMSO	Channel 3 EX BP 546/12 FT 580, LP 590
MDC	2.5 mM	10 μM	DMSO	Channel 2 G365 FT395, LP420
Hoechst 33342	1 g/L	1 mg/L	Water	Channel 1 G365FT395, LP420

8. Graded ethanol: Combine ethanol and distilled water in appropriate proportions to achieve the desired concentrations.
9. 2% uranyl acetate: Dissolve 2 g uranyl acetate in 100 mL ethanol.
10. 0.5% lead citrate: Fully dissolve 1.33 g lead nitrate and 1.76 g sodium citrate in 30 mL distilled water. Supplement with 8 mL 1 M NaOH to achieve a colorless and transparent solution. Meter with distilled water to a final volume of 50 mL and vigorously shake to ensure thorough mixing.

2.2 Equipment

1. Vacuum dryer.
2. Oven.
3. Horizontal mixer.
4. Automated cell counter.
5. Fluorescence microscope.
6. EM UC7 high-quality ultramicrotome.
7. JEM-1230 transmission electron microscope.
8. Grid.

3 Methods

3.1 Preparation of B. cinerea Conidial Samples

1. The *B. cinerea* strain B05.10 is preserved at −80°C prior to activation, which is subsequently incubated on PDA.
2. The activated strain is cultured in a light incubator at 25°C for 7–10 days to yield conidia.
3. Collect the conidia from the 10-day-old *B. cinerea* cultures on PDA and transfer them to sterile water.
4. Use sterile gauze to filter the conidial suspension and subsequently enumerate the count (*see* **Note 4**).
5. The conidial suspension is inoculated into PDB (10^6 CFU/mL) and subsequently cultured at 25°C for 6 h using a shaker (*see* **Note 5**).
6. Transfer 1 mL conidial suspension to a 1.5 mL microfuge tube, and subsequently spin it in a microcentrifuge to collect the conidia (Fig. 2).

3.2 Fluorescence Staining on the Conidia of B. cinerea

1. Rinse conidial samples with 0.1 M PBS buffer (pH 7.0) to eliminate residual contamination. The samples are resuspended in 0.5 mL PBS buffer at room temperature and subsequently collected by centrifugation at 10,000 × *g* for 30 s.
2. Rinsed samples are resuspended in 200 μL working solution of fluorescent dye. Subsequently, they are incubated in a water

bath at 30 °C for 0–30 min, according to the instructions for use of the fluorescent dye.

3. Wash the sample twice with PBS buffer to eliminate any residue from the fluorescent dye, followed by rinsing with 200 μL PBS buffer (*see* **Note 6**).
4. Transfer 5 μL samples to the center of the slide, then cover with a coverslip.
5. Observe under the fluorescence microscope (Leica DM2500, Germany) using the parameters for specific fluorescent probes (Table 1).
6. Capture images and subsequently process them using LAS AF Lite software equipped with a fluorescence microscope (Figs. 1B–D and 3).

3.3 Ultrastructural Observation of B. cinerea

1. Fixation: The samples (collected as described in Sect. 3.1) are treated with 1 mL of pre-cooled fixation solution 1, then vacuumed for 30 min, and subsequently stored at 4 °C for 24 h. Wash the fixed samples four times using cacodylate buffer (0.1 M, pH 7.2). The conidia are collected by a procedure as

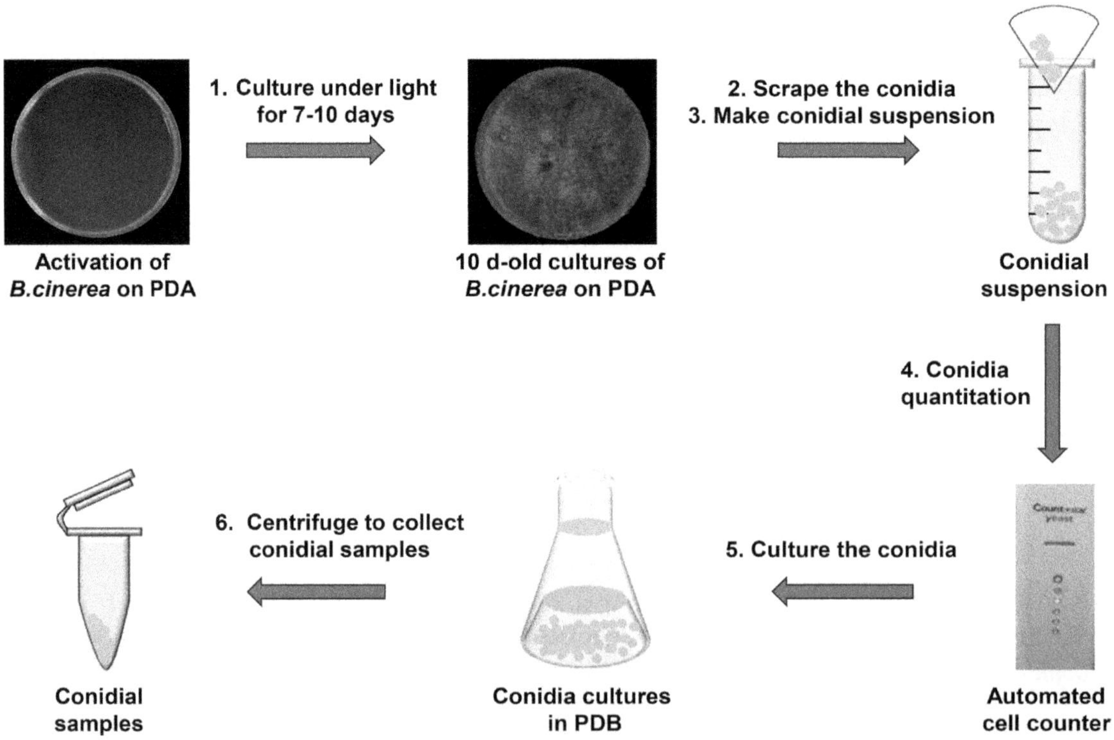

Fig. 2 Preparation process of *B. cinerea* samples

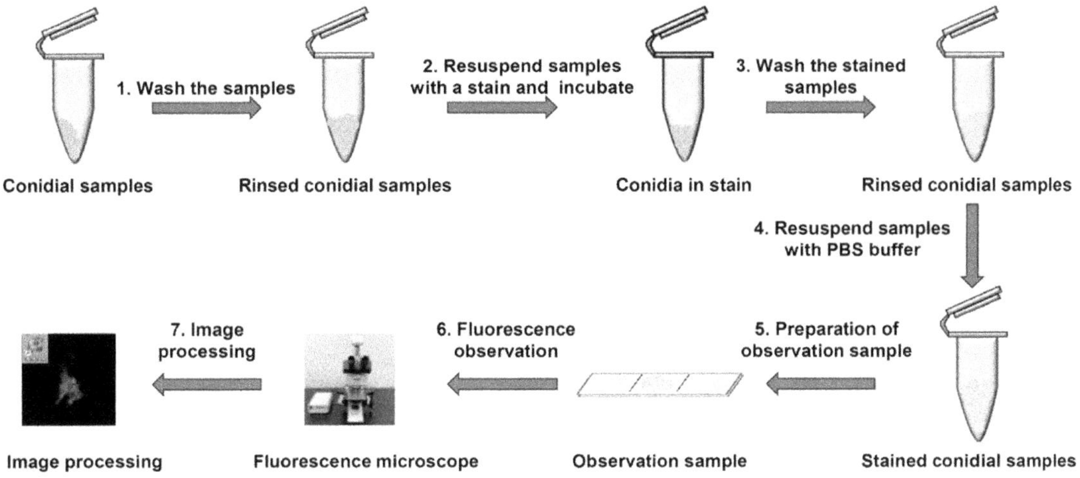

Fig. 3 Fluorescence staining procedures for *B. cinerea* conidia

follows: resuspended in 0.5 mL washing buffer, then kept on ice for 20 min, and subsequently collected by centrifugation at 3000 × *g* for 2 min. Add 0.3 mL fixation solution 2 and leave the samples at room temperature for 20 min. After the post-fixation process, rinse the samples four times with water.

2. Dehydration: Wash the fixed samples successively in 30%, 50%, 70%, 80%, 90% ethanol for 20 min each time. Then wash in 100% ethanol for 20 min twice (*see* **Note 7**).
3. Infiltration: Infiltrate samples successively in a mixture of acetone and Spurr low-viscosity embedding agent with different volume ratios (2:1, 1:1, 1:2) for 2 h each time (*see* **Note 8**).
4. Embedding: Soak samples in pure Spurr low-viscosity embedding agent twice for 8 h each.
5. Polymerization: Keep the embedded samples in a 60 °C oven for more than 24 h.
6. Sectioning: The samples were sectioned using a high-quality ultramicrotome (LEICA EM UC7), and the ultrathin sections were collected on the grid (*see* **Note 9**).
7. Staining: Stain the sections with 2% uranyl acetate and 0.5% lead citrate (*see* **Note 10**).
8. Observation: Capture images under a transmission electron microscope (JEM-1230, JEOL, Japan) (Figs. 1A and 4).

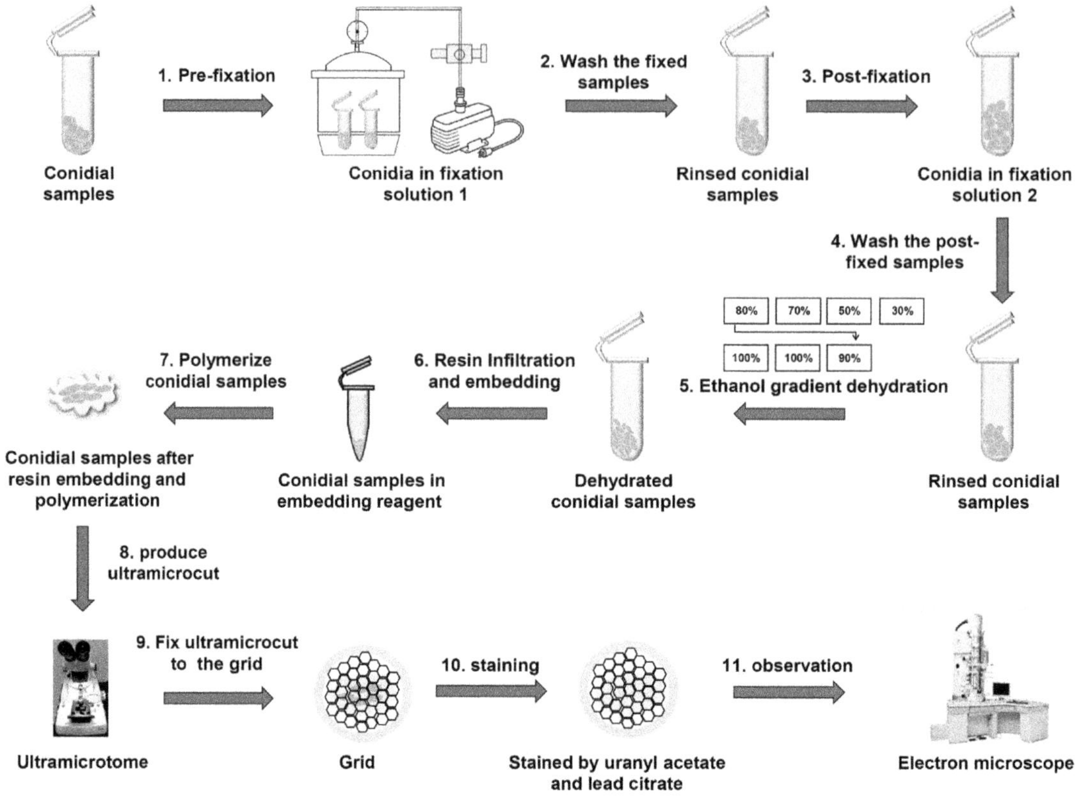

Fig. 4 Preparation and observation of ultrathin sections of *B. cinerea*

4 Notes

1. Most fluorescent dyes are insoluble in water; therefore, they should be dissolved using the solvent listed in the table. Given that fluorescent dyes are prone to decomposition under light exposure, it is crucial to ensure that they are not exposed to light during their preparation, storage, and application.
2. Although phosphate-buffered saline (PBS) is frequently used as the buffer for fixing biological samples, the fixation efficacy of sodium cacodylate buffer is better than PBS when applied to *B. cinerea* conidia.
3. The samples can also be post-fixed using fixation solution 2 (0.1 M sodium cacodylate buffer containing 0.5% OsO_4 and 2.5% $K_2Cr_2O_7$), which should be performed on ice for 90 min.
4. Conidia quantitation can be achieved using either the automated cell counter instrument or a blood cell counter plate by microscopic examination.

5. For the tests of exogenously applied agents, agents can be directly added to the PDB at respective concentrations at the beginning of culture.
6. To prevent rapid fluorescence quenching, the stained conidia should be kept on ice.
7. Here, ethanol exhibits a higher degree of dehydrating properties as compared to acetone.
8. The permeation and polymerization of Spurr low-viscosity embedding agent are stronger than that those of LR white.
9. The quality of ultrathin sections significantly influences subsequent observations. Consequently, this procedure necessitates considerable expertise and is ideally conducted by technical professionals.
10. The contrast of the sample can be improved through post-staining with lead citrate and uranyl acetate.

Acknowledgments

This work was supported by the National Natural Science Foundation of China (32372777) and the Beijing Municipal Natural Science Foundation (5242019, 6252024).

References

1. Dean R, Van Kan JA, Pretorius ZA, Hammond-Kosack KE, Di Pietro A, Spanu PD, Rudd JJ, Dickman M, Kahmann R, Ellis J, Foster GD (2012) The top 10 fungal pathogens in molecular plant pathology. Mol Plant Pathol 13:14–430
2. Zhang ZQ, Chen T, Li BQ, Qin GZ, Tian SP (2021) Molecular mechanism of pathogenesis of postharvest pathogenic fungi and control strategy in fruits: progress and prospect. Mol Hortic 1:2
3. Chen T, Zhang ZQ, Chen Y, Li BQ, Tian SP (2023) *Botrytis cinerea*. Curr Biol 33(11): R460–R462
4. Tian SP, Torres R, Ballester AR, Li BQ, Vilanova L, Gonzalez-Candelas L (2016) Molecular aspects in pathogen-fruit interactions: virulence and resistance. Postharvest Biol Technol 122:11–21
5. Wang GB, Zhao HN, Zou J, Liang WX, Zhao ZJ, Li DL (2024) Role of BcSfb3, the subunit of COPII vesicles, in fungal development and pathogenicity, ER-phagy and autophagy in the gray mold fungus *Botrytis cinerea*. Int J Biol Macromol 263:130379
6. Chen DG, Zhang ZQ, Chen Y, Li BQ, Chen T, Tian SP (2023) Transcriptional landscape of pathogen-responsive lncRNAs in tomato unveils the role of hydrolase encoding genes in response to *Botrytis cinerea* invasion. Plant Cell Environ 47:651–663
7. Ma DY, Cui XM, Zhang ZQ, Li BQ, Xu Y, Tian S, Chen T (2020) Honokiol suppresses mycelial growth and reduces virulence of *Botrytis cinerea* by inducing autophagic activities and apoptosis. Food Microbiol 88:103411
8. Huang XH, Liu W, Dong FQ, Xu Y, Tian SP, Chen T (2023) *Sapindus mukorossi* saponins inhibit gray mold on strawberry fruit by impairing membrane integrity and organellar homeostasis of *Botrytis cinerea*. Postharvest Biol Technol 107:112594

Check for updates

Chapter 6

Cell Biology of *Botrytis*: Live-Cell Imaging Techniques and Protein Localization Protocols

H. M. Suraj, André Fleissner and Jan A. L. van Kan

Abstract

Botrytis cinerea is a widely studied model organism for plant–fungal interactions. This chapter provides simple and easy protocols for live-cell imaging techniques and subcellular protein localization studies in *B. cinerea* in vitro. Using a simple microscopy setup, we describe methods for imaging spore germlings and hyphae and for studying protein dynamics in response to the antifungal compound α-tomatine. The protocols highlight the application of fluorescence microscopy to visualize sterol distribution and protein localization patterns, enabling deeper insights into fungal cell biology.

Keywords *Botrytis cinerea*, Microscopy, Live-cell imaging, Protein localization, Staining

1 Introduction

Botrytis cinerea serves as an excellent system for studying fungal cell biology and host–pathogen interactions. Understanding the cellular dynamics of *B. cinerea*, such as membrane repair, is critical for unraveling its pathogenic strategies and identifying potential antifungal targets. This chapter outlines detailed protocols for live-cell imaging of *B. cinerea*, with a case study on how to perform sterol staining using Filipin III and the localization of membrane repair factors using GFP-tagged proteins. By combining advanced microscopy techniques with targeted staining and subcellular protein localization studies, these protocols are valuable tools for investigating the cellular biology of *B. cinerea*.

2 Materials

2.1 Media

Liquid Minimal Medium (MM) for microscopy is prepared using the following recipe [1]:

Luca Nerva and Walter Chitarra (eds.), *Botrytis cinerea*, Methods in Molecular Biology, vol. 3076,
https://doi.org/10.1007/978-1-0716-5551-1_6,

Component	Quantity
Sucrose	20 g
Vogel's stock solution (50×)	20 mL
H_2O	1000 mL

2.2 Stock Solutions

2.2.1 Vogel's Stock Solutions (50×)

Component	Quantity
Sodium citrate	125 g
KH_2PO_4	250 g
NH_4NO_3	100 g
$MgSO_4 \times 7\,H_2O$	10 g
$CaCl_2 \times 2\,H_2O$	5 g
Chloroform (optional, only to keep sterile)	5 mL
Trace elements stock solution	5 mL
Biotin stock (50 mg in 1 L of 50% EtOH)	2.5 mL
ddH_2O	755 mL

2.2.2 Trace Elements Stock Solutions

Component	Quantity
Citric acid	5 g
$ZnSO_4 \times 7\,H_2O$	5 g
$Fe(NH_4)_2(SO_4)_2 \times 6\,H_2O$	1 g
$CuSO_4 \times 5\,H_2O$	0.25 g
$MnSO_4 \times H_2O$	0.05 g
Boric acid	0.05 g
$Na_2MoO_4 \times 2\,H_2O$	0.05 g
ddH_2O	100 mL

2.2.3 Stock Solution of Antifungal Compound (α-Tomatine)

Based on the polarity of the antifungal compound, decide which solvent is required to dissolve the compound to make a stock solution. Here, we consider α-tomatine as the antifungal plant metabolite.

1. α-Tomatine (20 mg/mL stock): 20 mg/mL of α-tomatine (T3636, TCI Europe) is dissolved in 75% MeOH containing 0.1% formic acid.
2. *Working stock of tomatine (1 mg/mL): prepared by taking* 6 μL of a 20 mg/mL tomatine stock solution and diluting it in 114 μL

of Minimal Medium (MM). Mix thoroughly to achieve the desired final concentration.

2.2.4 Staining Solutions

- Filipin III (10 mg/mL) stock solution: 10 mg of Filipin III (F4767, Sigma-Aldrich) dissolved in 1 mL DMSO.
- FM™ 4-64 Dye (1 mM) stock solution: 0.60751 mg of FM™ 4-64 (T13320, Invitrogen) dissolved in 1 mL DMSO.
- Working solution (100 μg/mL) Filipin III: 10 μL of 10 mg/mL stock in 990 μL dH_20.
- Working solution (10 μM) FM™ 4-64 Dye: 10 μL of 1 mM stock in 990 μL dH_20.

2.3 Imaging Setup Used

- Inverted Microscope: Zeiss Observer 2.1 with Nomarski optics and Plan-Neofluar 100×/1.30 oil immersion objective.
- Fluorescence Source: LED (CoolLED pE4000).
- Camera: PCO Edge 5.5 Gold (16 bit).
- ibidi eight-well μ-slides (Cat. No. 80826) to grow *Botrytis cinerea*.

3 Methods

3.1 Microscopy Technique for Sterol Staining Using Filipin III in *Botrytis*

Sterol staining using Filipin III is a well-established method to visualize sterol-rich regions in fungal cells, such as yeast cells, spore germlings, or hyphae [2]. The following protocol outlines the step-by-step procedure for preparing and imaging *Botrytis cinerea* using fluorescence microscopy (Fig. 1) .

1. **Preparation of Spore Suspension:** Under a clean bench, take a freshly harvested spore stock solution (10^6 spores/mL) and add 5 μL of *Botrytis cinerea* spore suspension to each well of an ibidi eight-well μ-slide. Spread the suspension carefully across the well to ensure even distribution. Allow spores to settle at the bottom for 2 min.
2. **Incubation in Minimal Medium**: Add 150 μL of liquid Vogel's Minimal Medium (MM) to each well. Secure the slide lids and incubate the slides at 20°C. For imaging germlings, incubate for 5–6 h. For longer septate hyphae, incubate for >12 h.
3. **Microscopy Setup**: Transfer the slides to an inverted microscope equipped with a 100× oil-immersion objective. Begin imaging individual germlings to capture preliminary observations.

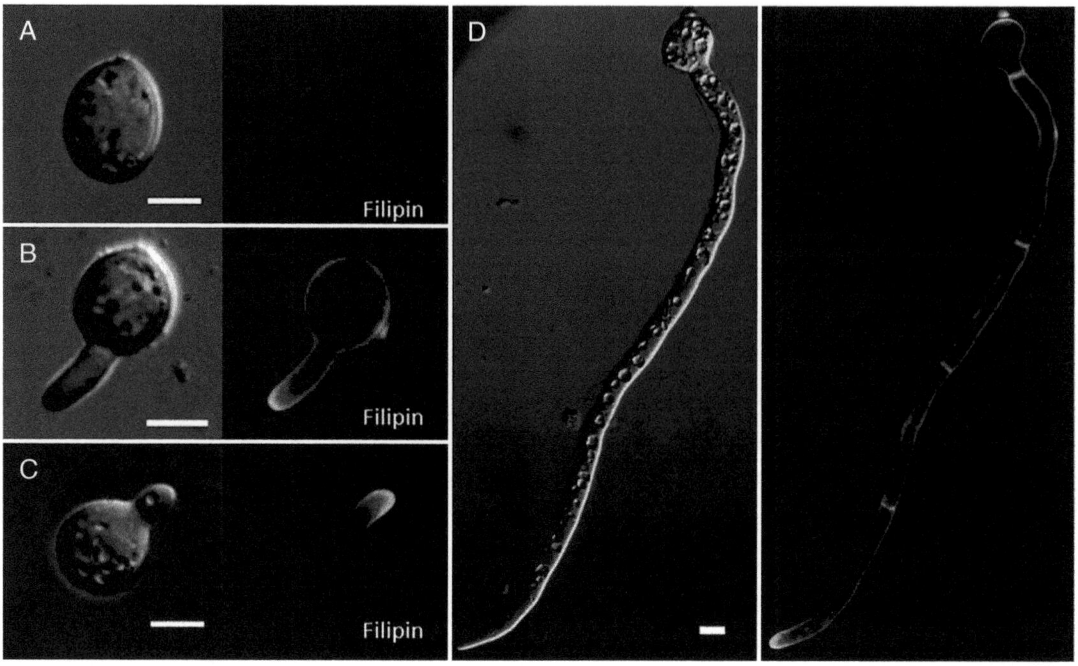

Fig. 1 Filipin III staining of spores, germlings and hyphae of *B. cinerea* (B05.10). Sterols are enriched at the tip of the germlings, hyphae, and septa. The scale bar indicates a distance of 5 μm for (**A–C**) and 2.5 μm for (**D**)

4. **Capture Bright-Field Image**: Before staining, take a bright-field/DIC image to visualize germling morphology.
5. **Staining Preparation**: As germlings adhere to the slide surface, carefully remove 150 μL of liquid MM from each well without disturbing the germlings. This step facilitates efficient staining with Filipin III.
6. **Add Filipin III Solution**: Add 100–150 μL of Filipin III working solution (100 μg/mL) to each well. Ensure even coverage of germlings and incubate slides in the dark for 20 min to avoid photobleaching.
7. **Fluorescence Imaging:** Using fluorescence microscopy, observe the stained germlings with a DAPI filter (excitation at 360 nm and emission at 480 nm). Sterol-enriched regions, such as hyphal tips and septa, will fluoresce, providing insights into sterol localization.

3.2 Microscopy Technique for Protein Localization Followed by Subsequent Sterol Staining Using Filipin III

This section describes live-cell imaging and in vitro protein localization in *Botrytis cinerea* based on a case study by [3]. This protocol utilizes two strains of *B. cinerea*: B05.10 OE-BcPEF1-GFP and B05.10 OE-BcGT28-GFP, which express membrane repair proteins tagged with GFP for visualization. These proteins, BcPEF1 and BcGT28, localize to damaged membrane sites after exposure to the membrane-permeating compound α-tomatine. Despite their shared role in membrane repair, the proteins exhibit distinct localization patterns, reflecting differences in biological functions (Figs. 2 and 3). Below are the step-by-step instructions for imaging protein localization of PEF1-GFP and GT28-GFP after treatment with antifungal metabolite α-tomatine, followed by subsequent sterol staining with Filipin III.

1. **Preparation of Spore Suspension:** Under a clean bench, add 5 μL of *Botrytis cinerea* spore suspension (10^6 spores/mL) into each well of an ibidi eight-well μ-slide. Carefully spread the spore suspension evenly across the bottom of the wells and allow the spores to settle for 2 min.
2. **Incubation in Minimal Medium:** Add 150 μL of liquid Vogel's Minimal Medium (MM) to each well. Close the slide lids securely and incubate the slides at 20°C for 5–6 h to allow germination of the spores for imaging germlings.
3. **Microscopy Setup:** Place the slides on an inverted microscope equipped with a 100× oil-immersion objective. Begin imaging individual germlings.
4. **Capture Bright-Field Image**: Before treatment, take a bright-field/DIC image at time zero (0′) to record the untreated state of the germlings.
5. **Treatment with α-Tomatine:** Since the germlings adhere to the slide surface, gently remove 150 μL of liquid MM from each well without disturbing the germlings. Add 100 μL of α-tomatine working solution (1 mg/mL) dissolved in liquid MM.
6. **Protein Localization Imaging:** After 10–20 min, visualize the changes in the localization of BcPEF1-GFP or BcGT28-GFP using a GFP fluorescence filter (excitation at 488 nm and emission at 510 nm). Capture images at different time intervals to observe the protein localization dynamics (Figs. 2 and 3).
7. **Staining with Filipin III:** After examining the protein dynamics, the cells can be further stained with either a live-cell or dead-cell stain, enabling applications such as co-localization studies. Here, we stain the germlings for sterol visualization. So, carefully remove 100 μL of liquid α-tomatine solution from the wells without disturbing the germlings.
8. **Add Filipin III Solution:** Add 100–150 μL of Filipin III working solution (100 μg/mL) to each well. Incubate the slides in the dark for 20 min to allow proper staining of sterols.
9. **Fluorescence Imaging:** Image the stained germlings using fluorescence microscopy with a DAPI filter (excitation at 360 nm, emission at 480 nm). This step enables visualization of sterol enrichment at specific cellular locations (Figs. 2 and 3).

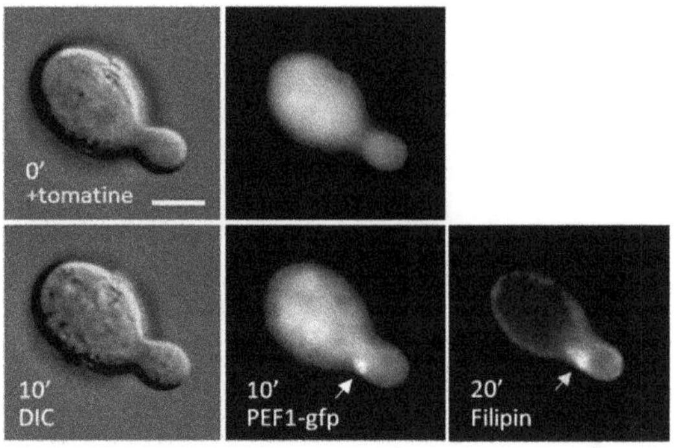

Fig. 2 Strain OE-BcPEF1-GFP was treated with α-tomatine at time 0, and imaged after 10 min. Subsequently, the strain was stained with Filipin III and photographed after 20 min. The sterol patch co-localized with the site of BcPEF1-GFP accumulation. Scale bar represents 5 μm

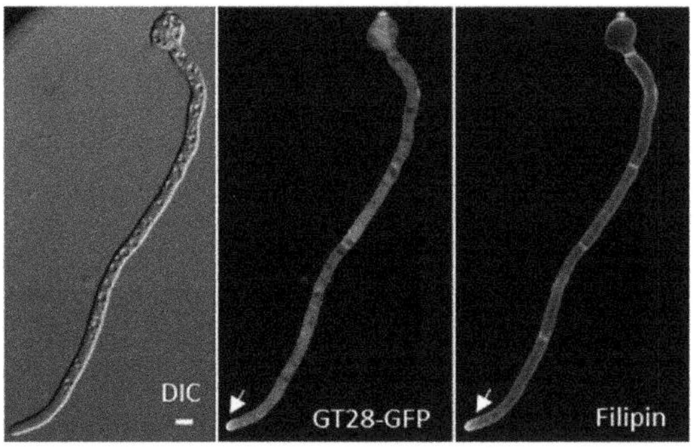

Fig. 3 Strain OE-BcGT28-GFP was treated with α-tomatine and imaged after 10 min. Subsequently, it was stained with Filipin III and photographed after 20 min. BcGT28-GFP localized to the hyphal tip, where sterols were also enriched. Scale bar represents 2.5 μm

4 Notes

1. Vogel's MM for microscopy is colorless and has no background fluorescence [1].
2. In steps 3.1.2 and 3.2.1, we use liquid Vogel's MM to inoculate the fungal spores instead of agar MM because germlings penetrate the agar, which hampers imaging of the germinating tips. Using the ibidi μ-slides with liquid medium allows to visualize the whole germlings as they stick to the surface.
3. Starting spore densities are crucial for optimum germination. At very high densities, spores will not germinate. Starting with lower densities and spreading spores before adding liquid medium ensures better distribution over the slide surface.

4. Other fluorescent dyes, such as FM4-64 (a live membrane stain), can be combined with the imaging of GFP-labeled proteins (step 3.2.6). FM4-64 has been used [4] to visualize plasma membrane and intracellular membrane structures. The dye binds to the plasma membrane and is internalized via endocytosis, labeling endosomes and other intracellular vesicles. Short incubation times result in staining of the plasma membrane only, while extended incubation also highlights intracellular structures. Observing membrane and vesicular dynamics using FM4-64 can complement protein localization studies and provide deeper insights into protein function.

Acknowledgments

H.M. Suraj acknowledges TKI Graduate School Green Top Sectors (grant LWV21.291, TU-2021-20), The Netherlands, for funding his project.

References

1. Vogel HJ (1956) A convenient growth medium for *Neurospora crassa*. Microb Genet Bull 13:42–47
2. Van Leeuwen MR, Smant W, de Boer W, Dijksterhuis J (2008) Filipin is a reliable *in situ* marker of ergosterol in the plasma membrane of germinating conidia (spores) of *Penicillium discolor* and stains intensively at the site of germ tube formation. J Microbiol Methods 74:64–73. https://doi.org/10.1016/j.mimet.2008.04.001
3. You Y, Suraj HM, Matz L, Herrera Valderrama AL, Ruigrok P, Shi-Kunne X, Pieterse FPJ, Oostlander A, Beenen HG, Chavarro-Carrero EA, Qin S, Verstappen FWA, Kappers IF, Fleißner A, van Kan JAL (2024) *Botrytis cinerea* combines four molecular strategies to tolerate membrane-permeating plant compounds and to increase virulence. Nat Commun 15(1):6448. https://doi.org/10.1038/s41467-024-50748-5
4. Hickey PC, Swift SR, Roca MG, Read ND (2004) Live-cell imaging of filamentous fungi using vital fluorescent dyes and confocal microscopy. In: Methods in microbiology, vol 34. Academic Press, pp 63–87. https://doi.org/10.1016/S0580-9517(04)34003-1

Check for updates

Chapter 7

Methods for Inoculating *Botrytis cinerea* on Leaves and Fruit

Daoguo Chen, Tong Chen, Boqiang Li, Yong Chen and Shiping Tian

Abstract

Being a generalist pathogen with a broad host range, *Botrytis cinerea* causes tremendous postharvest loss and food waste, while the incidence of gray mold is contingent upon the pathogenicity of *B. cinerea* and the resistance of host plants. It can thus be anticipated that the goals for controlling gray mold can be achieved by either inhibiting the pathogenicity of *B. cinerea* or enhancing the resistance of host plants. To assess the pathogenicity of pathogens or the defense responses of hosts, a standardized inoculation protocol is indispensable. Notably, specific controlled conditions are also important requisites. In this chapter, we primarily summarized a standard inoculation method and some key notes for the experimental procedures.

Keywords *Botrytis cinerea*, Inoculation method, Fruit, Leaves

1 Introduction

Botrytis cinerea, a necrotic fungal pathogen, induces severe gray mold disease in both preharvest and postharvest stages [1] (Fig. 1). A bulk of studies have been carried out to investigate the mechanisms underlying the pathogenesis of gray mold with emphasis on either host plants or *B. cinerea*. Genes regulating host defense responses have been identified, such as *SlMYC2*, *SlVPE3*, and *SlFERL* [2–4]. Meanwhile, genes contributing to the pathogenicity of *B. cinerea* have also been examined, such as *BcAra1*, *Bcsnf1*, *BcNoxR*, *BcRho3*, *BcAQP8*, *Bcsas1*, and *BcactA* [5–7].

To dissect the roles of defense-related genes in plants and virulence factors of *B. cinerea*, and further evaluate the inhibitory effect of active substances on *B. cinerea*, a standard inoculation protocol and strictly controllable conditions are essential [8]. Furthermore, examinations on gene functions, such as omics analyses, also require high-quality data following consistent inoculation processes and sampling standards [9–12]. Therefore, this chapter aims to provide an inoculation protocol to ensure the stability of the phenotype of *B. cinerea* on plant tissues. This protocol consists of five main steps:

Luca Nerva and Walter Chitarra (eds.), *Botrytis cinerea*, Methods in Molecular Biology, vol. 3076,
https://doi.org/10.1007/978-1-0716-5551-1_7,

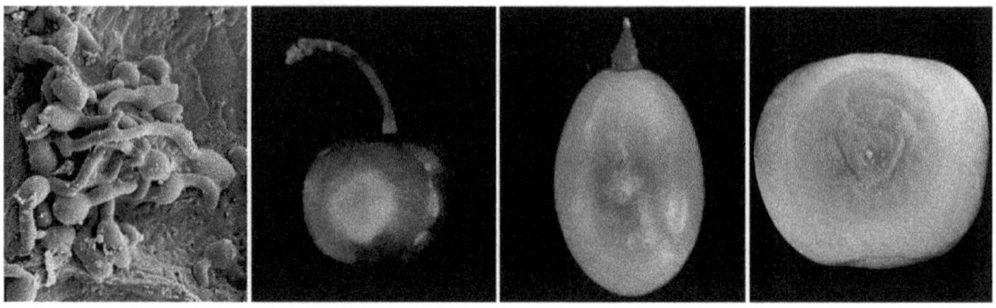

Fig. 1 *Botrytis cinerea* and gray mold on fruit

(i) culture of *B. cinerea*, (ii) preparation of the spore suspension, (iii) fruit surface sterilization, (iv) inoculation on fruit and leaves, and (v) phenotypic analysis. Additionally, we summarized some notes for the inoculation process on plant tissues.

2 Materials

2.1 Plant and Fungal Pathogen Materials

1. *B. cinerea* strain B05.10.
2. Tomato (*Solanum lycopersicum* cv. Micro-Tom) plants, which were cultivated in a greenhouse under conditions of 16 h of light and 8 h of darkness at 23°C.
3. Tomato (*S. lycopersicum* cv. Chuan-Hong) fruit at the mature green stage (MG), which were purchased from a local supermarket.

2.2 Equipment

1. Electric cooker.
2. Sterile gauze.
3. Centrifuge tubes of 1.5, 10, and 50 mL.
4. Surgical scalpel blades.
5. Autoclave.
6. High-speed centrifuge.
7. Clean bench.
8. Petri dishes of 9 cm diameter.
9. Alcohol burner.
10. Miracloth.
11. Automated cell counter.
12. Constant-temperature incubator.
13. Sterile pipette tips (200 μL).
14. Ruler or vernier caliper.

2.3 Required Reagents

1. Potato dextrose broth (PDB): Peel and cut potatoes into pieces (2 cm × 2 cm × 2 cm). Place 200–230 g of potato pieces into 1000 mL of distilled water and boil for 30 min. Filter the mixture through an eight-layer sterile gauze and transfer the filtrate into 50 mL centrifuge tubes. Centrifuge for 30 min at 10,000 × *g* and

retain the supernatant. Add 20 g of glucose and adjust the total volume to 1000 mL with distilled water. Autoclave at 121°C for 20 min.
2. Potato dextrose agar (PDA): The preparation of PDA follows the same procedure as that for PDB, except the addition of 15 g agar.
3. Sodium hypochlorite.
4. Sterile distilled water.

3 Methods

As depicted in Fig. 2, spore suspensions and mycelial plugs of *B. cinerea* are utilized for inoculating leaves and fruit.

3.1 Culture of *B. cinerea*

1. Establish a sterile work area on the clean bench.
2. Inoculate *B. cinerea* spore suspension (stored at −80°C with 20% glycerol) into the center of a PDA plate using a 10 μL pipette.
3. Culture *B. cinerea* in a constant-temperature incubator (22°C) with exposure to light for 7–14 days.

3.2 Preparation of Spore Suspension

1. Wash *B. cinerea* cultures with sterile water.
2. Filter the *B. cinerea* spores through two layers of Miracloth.
3. Determine the concentration of the *B. cinerea* spore suspension using an automated cell counter.

3.3 Fruit Surface Sterilization

1. Sterilize the fruit surface in 2% sodium hypochlorite for 2 min.
2. Rinse the sterilized fruit twice with sterile water.

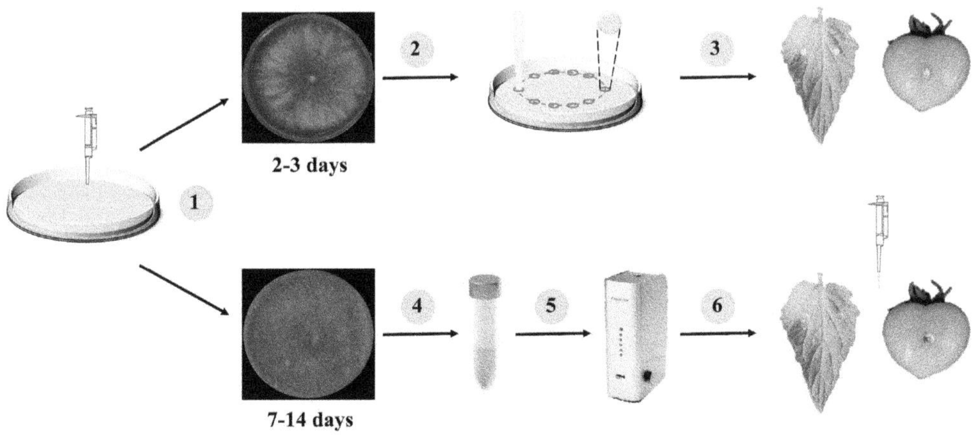

Fig. 2 Workflow of *B. cinerea* inoculation on fruit and leaves. (1) Inoculate *B. cinerea* spores onto the PDA. (2) Prepare mycelial plugs of *B. cinerea*. (3) Add mycelial plugs to leaves or fruit wounds, with one side of the mycelium remaining in contact with the plant tissue. (4) Prepare *B. cinerea* spore suspension. (5) Verify the concentration of the spore suspension and dilute it to the required concentration. (6) Add spore suspension to leaves or fruit wounds

3. After air-drying, stab the fruit at the equatorial zone using a sterile needle (2–3 mm wide × 2–3 mm deep).
4. Keep the fruit at room temperature for 30 min, then wipe off the fluid from the wound before inoculation.

3.4 Inoculation on Leaves and Fruit

3.4.1 Inoculate B. cinerea Spore Suspension on Leaves and Fruit

1. Dilute the spore suspension (prepared according to Sect. 3.1) to an appropriate concentration (2×10^5 or 2×10^6 spores/mL) using sterile water (for fruit inoculation), 1/4, 1/2 PDB, or PDB (for leaf inoculation).
2. Apply 500–20,000 spores to fruit wounds [10,13–15]. Apply 1000 spores to leaves or vegetables [2, 10].
3. Incubate inoculated leaves and fruits at room temperature (approximately 22°C–25°C) with high humidity (90%–95% relative humidity, RH).

3.4.2 Inoculate B. cinerea Mycelia onto Leaves and Fruit

1. Cut off the top part of pipette tips (200 μL) with a blade.
2. Sterilize pipette tips at 121°C for 20 min.
3. Select *B. cinerea* cultured for 2–3 days and punch holes (about 2 mm in diameter) using a pipette tip.
4. Add *B. cinerea* mycelial plugs to leaves or fruit wounds.

3.5 Phenotypic Analysis

After inoculation for a designated period, measure the lesion diameter using a ruler or vernier caliper (magnified intersections method) [16].

The quantity of inoculated spores and the status of plant materials significantly affect the pathogenesis of gray mold. Accordingly, different observation times are required for various inoculation conditions and plant materials. As shown in Fig. 3, 1×10^3 spores were

Fig. 3 Phenotypic observation of tomato leaves inoculated with *B. cinerea* spores. The degree of gray mold incidence in leaves

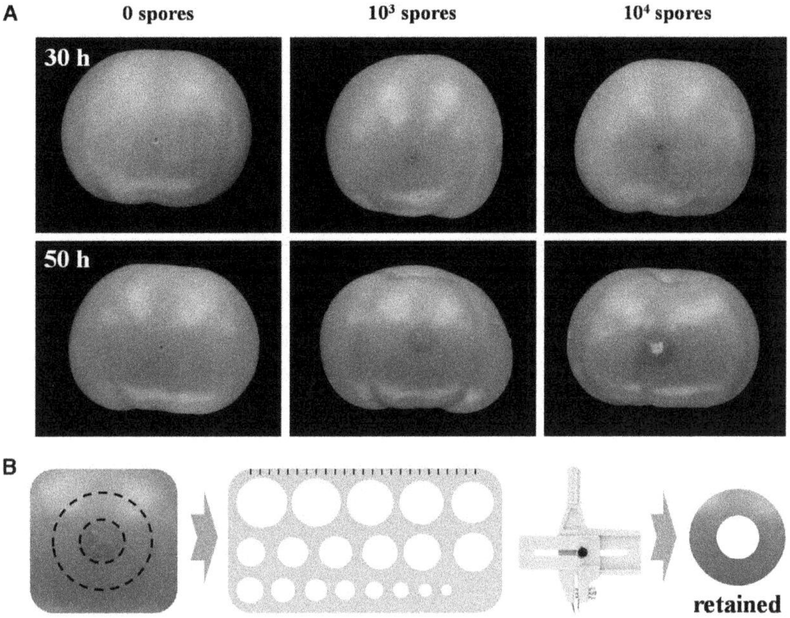

Fig. 4 Phenotypic observation and sample collection of inoculated tomato fruit. (**A**) Tomato fruit inoculated with various quantities of *B. cinerea* spores. (**B**) Schematic diagram of sample collection. Following *B. cinerea* inoculation, the lesion area was removed, and fruit tissues (about 3 mm from the boundary of the lesion) were reserved for further experiments

inoculated on tomato leaves. The results showed that necrotic lesions appeared at 24 h post-inoculation (hpi), and thus the lesion sizes were recorded at 36, 48, 60, and 72 hpi, respectively. Moreover, 48 hpi was chosen for lesion diameter measurement after inoculating 1×10^3 spores on tomato leaves [10]. Varied pathogenesis was observed in *B. cinerea*-inoculated tomato fruit due to differing spore inoculum quantities (Fig. 4). Additionally, after inoculating strawberry, grape, sweet cherry, and apple fruit with 1×10^3 *B. cinerea* spores, lesion diameter was measured at 48 and 72 hpi [17–19]. To analyze the change in transcript level in *B. cinerea*-inoculated fruit, Chen et al. removed the lesion area and reserved fruit tissues (about 3 mm from the boundary of the lesion), ensuring the accuracy of sample collection (Fig. 4b) [8].

4 Notes

1. To minimize the risk of contamination of other pathogens, inoculation should be conducted in a sterile environment whenever possible.
2. Cultures of *B. cinerea* should be exposed to light to facilitate spore production.

3. Before measuring the spore concentration, the spore suspension should be vigorously vortexed, and measurements should be taken in triplicate to ensure accuracy.
4. As depicted in Fig. 4A, to guarantee consistent spore inoculation quantities, the spore suspension should be fully mixed, and spores should be collected from the center of the suspension during inoculation.
5. To maintain spore viability, the collected spore suspension should be used for inoculation as soon as possible (within 2 h).
6. To ensure consistent inoculation of mycelial plugs, the punching position should be located on the same concentric circles (Fig. 5B).
7. Select plant materials that are consistent in development, size, and appearance to ensure a uniform phenotype.
8. To ensure that the spores are inoculated into the wound in the following step, the fruit should be kept for a short period (about 30 min for tomato fruits) after punching, and the excessive fluid from the wound is carefully wiped off with a paper towel.
9. To ensure a stable incidence rate during leaf inoculation, it is necessary to provide certain nutrients. Therefore, the spore suspension was suspended using 1/4 PDB, 1/2 PDB, or PDB.

Fig. 5 Key considerations during *B. cinerea* inoculation on leaves and fruit

10. To maintain a stable incidence rate, an experimental environment with high RH is essential. As shown in Fig. 5C, the inoculation box was wrapped with a fresh-keeping bag, with wet paper inside to maintain constant RH. In contrast, *B. cinerea*-inoculated leaves can be kept on ilter paper pre-soaked with sterile water or 0.8% agar (Fig. 5D).
11. When conducting physiological index detection, gene expression analysis, and omics analysis, the samples from the same developmental stage and experimental conditions, as well as a unified standard, should be used (Fig. 4B).

References

1. Chen T, Zhang Z, Chen Y, Li BQ, Tian SP (2023) *Botrytis cinerea*. Curr Biol 33(11):R460–R462
2. Ji DC, Liu W, Cui XM, Liu K, Liu YH, Huang XH, Li BQ, Qin GZ, Chen T, Tian SP (2023) A receptor-like kinase SlFERL mediates immune responses of tomato to *Botrytis cinerea* by recognizing BcPG1 and fine-tuning MAPK signaling. New Phytol 240:1189–1201
3. Min DD, Li FJ, Cui XX, Zhou JX, Li JZ, Ai W, Shu P, Zhang XH, Li XA, Meng DM, Guo YY, Li J (2020) SlMYC2 are required for methyl jasmonate-induced tomato fruit resistance to *Botrytis cinerea*. Food Chem 310:125901
4. Wang WH, Cai JH, Wang PW, Tian SP, Qin GZ (2017) Post-transcriptional regulation of fruit ripening and disease resistance in tomato by the vacuolar protease SlVPE3. Genome Biol 18(1):47
5. Lengyel S, Rascle C, Poussereau N, Bruel C, Sella L, Choquer M, Favaron F (2022) Snf1 kinase differentially regulates *Botrytis cinerea* pathogenicity according to the plant host. Microorganisms 10(2):444
6. Nafisi M, Stranne M, Zhang L, van Kan JA, Sakuragi Y (2014) The endo-arabinanase BcAra1 is a novel host-specific virulence factor of the necrotic fungal phytopathogen *Botrytis cinerea*. Mol Plant Microbe Interact 27:781–792
7. Zhang ZQ, Chen Y, Li BQ, Chen T, Tian SP (2020) Reactive oxygen species: a generalist in regulating development and pathogenicity of phytopathogenic fungi. Comput Struct Biotechnol J 18:3344–3349
8. Chen DG, Chen T, Chen Y, Zhang ZQ, Li BQ, Tian SP (2023) Bio-source substances against postharvest diseases of fruits: mechanisms, applications and perspectives. Postharvest Biol Technol 198:112240
9. Blanco-Ulate B, Vincenti E, Powell ALT, Cantu D (2013) Tomato transcriptome and mutant analyses suggest a role for plant stress hormones in the interaction between fruit and *Botrytis cinerea*. Front Plant Sci 4:142
10. Chen DG, Zhang ZQ, Chen Y, Li BQ, Chen T, Tian SP (2024) Transcriptional landscape of pathogen-responsive lncRNAs in tomato unveils the role of hydrolase encoding genes in response to *Botrytis cinerea* invasion. Plant Cell Environ 47:651–663
11. Li H, Zhang ZQ, He C, Qin GZ, Tian SP (2016) Comparative proteomics reveals the potential targets of *BcNoxR*, a putative regulatory subunit of NADPH oxidase of *Botrytis cinerea*. Mol Plant Microbe Interact 29(12):990–1003
12. Liu K, Liu W, Huang XH, Liu YH, Cui XM, Zhang ZQ, Li BQ, El-Mogy MM, Tian SP, Chen T (2023) Identification of virulence-related proteins during *Botrytis cinerea* – fruit interaction at early phase. Postharvest Biol Technol 204:112443
13. He C, Zhang ZQ, Li BQ, Xu Y, Tian SP (2019) Effect of natamycin on *Botrytis cinerea* and *Penicillium expansum*—postharvest pathogens of grape berries and jujube fruit. Postharvest Biol Technol 151:134–141
14. Lin ZF, Alexander L, Hackett R, Grierson D (2008) LeCTR2, a CTR1-like protein kinase from tomato, plays a role in ethylene signalling, development and defence. Plant J 54:1083–1093
15. Silva CJ, Adaskaveg JA, Mesquida-Pesci SD, Ortega-Salazar IB, Pattathil S, Zhang L, Hahn MG, van Kan JAL, Cantu D, Powell ALT, Blanco-Ulate B (2023) *Botrytis cinerea* infection accelerates ripening and cell wall disassembly to promote disease in tomato fruit. Plant Physiol 191(1):575–590
16. Mc-Gonigle TP, Miller MH, Evans DG, Fairchild GL, Swan JA (1990) A new method which gives an objective measure of colonization of roots by vesicular-arbuscular mycorrhizal fungi. New Phytol 115:495–501

17. Cui XM, Ma DY, Liu XY, Zhang ZQ, Li BQ, Xu Y, Chen T, Tian SP (2021) Magnolol inhibits gray mold on postharvest fruit by inducing autophagic activity of *Botrytis cinerea*. Postharvest Biol Technol 180:111596
18. Li GJ, Chen Y, Zhang ZQ, Li BQ, Chen T, Tian SP (2022) 2,3-Butanedione suppresses gray mold of postharvest fruit by activating the autophagy of *Botrytis cinerea*. Postharvest Biol Technol 193:112057
19. Li GJ, Wang Y, Zhang ZQ, Chen Y, Tian SP (2022) Mushroom alcohol controls gray mold caused by *Botrytis cinerea* in harvested fruit via activating the genes involved in jasmonic acid signaling pathway. Postharvest Biol Technol 186:111843

Check for updates

Chapter 8

Protocol for the Extraction of High-Quality DNA from *Botrytis cinerea*

Chen Ya, Liu Yiqing and Jin Weibo

Abstract

Using *Botrytis cinerea* (ATCC 58025) as the material, genomic DNA was extracted using the SDS-CTAB method after liquid nitrogen freezing and extracted multiple times with organic reagents. Then the precipitate was washed with ethanol and dissolved in TE buffer. The cultured genomic DNA is high-yield, easy to extract, and has good purity, making it suitable for extracting from multiple samples at once. After experiments, the genomic DNA obtained by this method can be used for most basic molecular biology experiments such as PCR and DNA enzyme digestion.

Keywords ATCC 58025, Liquid nitrogen, TE buffer, Easy, High purity, SDS-CTAB

1 Introduction

Botrytis cinerea is a pathogenic fungus that can host plants such as tomatoes, strawberries, and flowers. Screening efficient and non-toxic biopesticides has become an urgent problem. Therefore, it is crucial to gain a deeper understanding of the genome of *Botrytis cinerea* and identify genes that can inhibit its proliferation and development. High-quality DNA must be obtained to conduct molecular-level research. The method used in this chapter is an improvement of the SDS-CTAB method, aiming to find a fast and efficient genomic DNA extraction method, obtain high-quality DNA suitable for molecular biology research, and provide material support for further prevention and control of gray mold.

The method in this chapter uses liquid nitrogen assistance to grind *Botrytis cinerea* and add it to DNA extraction buffer and 10% SDS solution for a water bath. Then, CTAB is added to continue the water bath. After centrifugation, the mixed reagents of phenol,

These authors contributed equally: Chen Ya and Liu Yiqing.

Luca Nerva and Walter Chitarra (eds.), *Botrytis cinerea*, Methods in Molecular Biology, vol. 3076,
https://doi.org/10.1007/978-1-0716-5551-1_8,

chloroform, and isoamyl alcohol, followed by chloroform and isoamyl alcohol are used for extraction, followed by isopropanol precipitation. Finally, the DNA precipitate is washed with ethanol and dissolved in TE buffer [1--4]. The DNA cultured by this method has high purity and good quality, which can be used for most basic experiments in analytical biology, such as double enzyme digestion, PCR, etc.

2 Materials

2.1 Reagents Setup

- DNA buffer: 1 M Tris-HCl 200 μL (pH 8.0), 4 M NaCl 1.5 mL, 0.5 mM EDTA 800 μL, add ddH_2O to bring the volume to 40 mL. The buffer can be stored at 4°C for about 1 year.
- 5 M NaCl: NaCl 14.61 g, add ddH_2O to dissolve and to bring the volume to 50 mL. This reagent can be stored at 18–27°C for 1 year.
- CTAB/NaCl buffer: CTAB 2.5 g, NaCl 1.025 g, add ddH_2O to dissolve and to bring the volume to 25 mL.
- 10% SDS: SDS 1.44 g, add ddH_2O to dissolve and to bring the volume to 50 mL.
- 70% Ethanol: Absolute ethanol 35 mL, add ddH_2O to bring the volume to 50 mL.
- Phenol:Chloroform:Isoamyl alcohol (25:24:1): Phenol 25 mL, chloroform 24 mL, isoamyl alcohol 1 mL. It is sealed with 100 mM Tris-HCl (pH 8.0) and stored in the dark at 4°C.

  CRITICAL: Phenol, chloroform, and isoamyl alcohol should be handled with protective gloves in a fume hood.
- Chloroform:Isoamyl alcohol (24:1): Chloroform 48 mL, isoamyl alcohol 2 mL. The reagent can be stored at 4°C for 1 year.

  CRITICAL: Chloroform and isoamyl alcohol should be handled with protective gloves in a fume hood.
- TE buffer: 1 M Tris-HCl 100 μL (pH 8.0), 0.5 mM EDTA 20 μL, add ddH_2O to bring the volume to 10 mL.
- 1% Agarose gel: Agarose 0.3 g, TAE buffer 30 ml. Boil the liquid until clear before pouring.
- Isopropanol.
- Liquid nitrogen.

2.2 Equipment

- Vortex mixer.
- 4°C or −20°C refrigerator.
- Sterilized mortar.
- Microcentrifuge tube.

- Electronic balance.
- Water bath.

2.3 Consumables

- 1.5 mL microcentrifuge tubes.
- 100 mL pipette and corresponding tips.
- 200 mL pipette and corresponding tips.
- 1000 mL pipette and corresponding tips.

3 Methods

1. Collect 20 mg of mycelium of *B. cinerea* from a 5-day-old PDA culture (Fig. 1) and store it at –80°C if the collected mycelium is not used for DNA extraction within 3 h.
2. Grind the mycelium in a sterilized mortar with liquid nitrogen and then transfer it to a 1.5 mL microcentrifuge tube.
3. Add 500 μL of DNA buffer to suspend the sample, vortex and mix well. Then add 50 μL of 10% SDS, mix well, and place in a water bath at 37°C for 1 h.
4. Add 75 μL of 5 M NaCl and 65 μL of CTAB/NaCl buffer, then place in a water bath at 65°C for 20 min.
5. Place the sample on ice or in a refrigerator to cool it to room temperature.
6. Add an equal volume (690 μL) of phenol:chloroform:isoamyl alcohol, mix well, and centrifuge at 10,000 × *g* for 10 min.

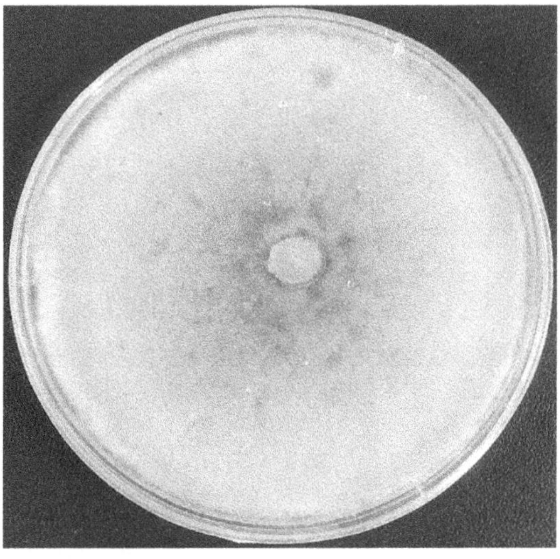

Fig. 1 *Botrytis cinerea* growing on a PDA plate for 5 days

7. Transfer the supernatant (500/600 μL) to a new 1.5 mL microcentrifuge tube, add equal volumes of chloroform:isoamyl alcohol, invert and mix well, and centrifuge at 10,000 × *g* for 10 min.
8. Transfer the supernatant, add 0.6 times the volume of isopropanol, invert and mix well, and centrifuge at 10,000 × *g* for 10 min.
9. Discard the supernatant, add 1 ml of 70% ethanol, invert and mix well, and centrifuge at 10,000 × *g* for 10 min.
10. Discard the supernatant, dry the sample for 5 min with the lid open, add 30 μL of TE buffer to dissolve it, use agarose gel electrophoresis to test the product, and store the product at −20°C for a long time.

4 Outputs

4.1 Detection of Genomic DNA by Agarose Gel Electrophoresis

Take 5 μL of genomic DNA extracted by the above method for agarose gel electrophoresis detection. The experimental results are shown in Fig. 2. The agarose electrophoresis results show that the genomic DNA extracted by the method has good integrity and a narrow band, indicating little degradation. The results from the third lane show that the DNA extracted by this method still performs well after being stored at −20°C for 3 months.

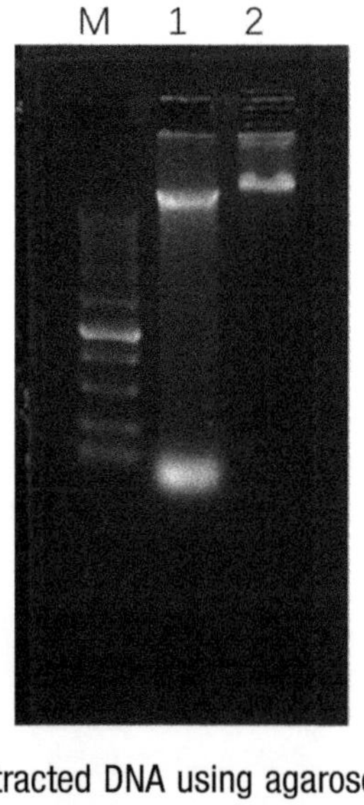

Fig. 2 Detection of the extracted DNA using agarose gel electrophoresis. Lane M: DL2000. Lane 1: Extracted DNA of *B. Cinerea* without RNase A digestion. Lane 2: Extracted DNA of *B. Cinerea* with RNase A digestion

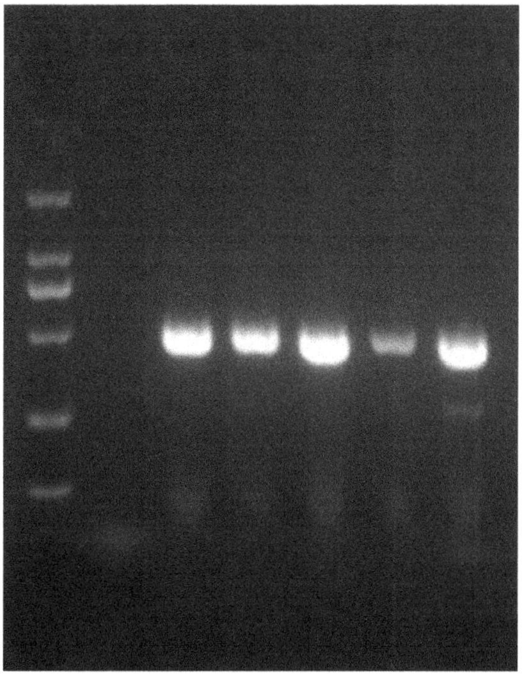

Fig. 3 The quality of the extracted DNA was further detected by PCR. Lane M: DL2000 DNA marker; Lane 1: The product of PCR amplification without adding extracted DNA (negative control); Lanes 2–6: PCR products amplified at five different temperatures, including 55°C, 56°C, 5 °C, 58°C, and 59°C

The concentration of the samples was measured by the NanoDrop-2000 instrument, and the results showed that the concentration is 2648.7 ± 2.0 ng/μL.

4.2 PCR Detection of Genomic DNA

The genomic DNA extracted by this method was subjected to temperature gradient PCR using the fungal ITS primers (Forward primer: 5′-TCCGTAGGTGAACCTGCGG-3′; Reverse primer: 5′-TCCTCCGCTTATTGATATGC-3′) [5], which were already available in the laboratory, and the results are shown in Fig. 3. The ITS gene corresponds to the login number OW983047.1 on the NCBI website, with a fragment length of 512 bp, which matches the corresponding bands in Fig. 3, proving that the amplified band is the ITS gene. The genomic DNA extracted by this method is suitable for PCR detection.

5 Notes and Troubleshooting

Problem 1
Mycelium is prone to splashing during grinding.

Potential solution

Wait until the liquid nitrogen has evaporated before grinding. Press the grinding rod firmly against the frozen *Botrytis cinerea*; do not hit the *Botrytis cinerea* too hard, as it will cause the mycelium to splash, resulting in waste and insufficient grinding.

Problem 2

The volume of gray mold powder is too small to be transferred to a 1.5 mL microcentrifuge tube directly.

Potential solution

Add 500 μL of DNA solution to the mortar directly, continue grinding, mix well, and then transfer to a microcentrifuge tube with a pipette. Before suction, mix the suspended liquid with a pipette and try to aspirate the solution containing *Botrytis cinerea* particles at the bottom.

Problem 3

CTAB/NaCl buffer is a colorless viscous solution, and it is difficult to accurately transfer a volume of solution with a pipette at room temperature.

Potential solution

CTAB/NaCl buffer can be heated in a 65°C water bath in advance to reduce its viscosity.

Problem 4

It is easy to collect the lower solution and cause contamination when absorbing the supernatant.

Potential solution

Transfer less supernatant to prevent contamination.

Acknowledgments

This work was supported by a grant from the Natural Science Foundation of China (grant no. 32172496).

References

1. Conlon BH, Schmidt S, Poulsen M, Shik JZ (2022) Orthogonal protocols for DNA extraction from filamentous fungi. STAR Protoc 3(1):101126
2. Keriö S, Terhonen E, LeBoldus J (2020) Safe DNA-extraction protocol suitable for studying tree-fungus interactions. Bio Protoc 10(11):3634
3. Yang Y, Zuzak K, Feng J (2016) An improved simple method for DNA extraction from fungal mycelia. Can J Plant Pathol 38(4):476–482
4. Zhang B, Lv S, Lin Y, Ma D, Zhang S, Xiao BJ (2006) Study on DNA extraction methods in fusants against *Botrytis Cinerea* Pers. on tomato. Biotechnology 16(5):32–34
5. Xu Y, Qi Y, Wen S (2020) Isolation and molecular identification of the pathogenic fungi of black spot of *Fritillaria thunbergii* Miq. J Zhejiang Chin Med Univ 44(2):111–118

Chapter 9

Strain Differentiation by SSR-PCR Using Capillary Sequencer and Agarose Gel Electrophoresis

Louis Backmann and Maren Scharfenberger-Schmerr

Abstract

Simple sequence repeat markers (SSRs) can be used in SSR-PCR for the differentiation and identification of organisms. Here, we describe two methods for strain differentiation using the example of *Botrytis cinerea*, differing in their complexity, by capillary sequencer and agarose gel electrophoresis. Nearly all analyzed strains could be distinguished by both methods and assigned a unique "fingerprint."

Keywords SSR-PCR, Capillary sequencer, Agarose gel electrophoresis, Strain differentiation, Fingerprinting

1 Introduction

Simple sequence repeat markers (SSRs) are typically specific primers, targeting conserved regions in a species. Due to slightly diverging lengths of the fragments between strains, using multiple SSRs will provide a "fingerprint" of every tested strain. This can be used to characterize various species [1–4]. There are multiple methods for downstream analysis of SSR-PCR, such as capillary sequencing, a very accurate and fast method to distinguish fragment sizes of up to 1 base pair [5, 6], or agarose gel electrophoresis, a method for visualizing the fragments on a gel, which is less accurate [6–8], providing distinguishing fragments up to 5–10 basepairs. Using a capillary sequencer is typically more complex and not always provided by the given lab equipment, whereas agarose gel electrophoresis is cheaper and often standard laboratory equipment. We were able to show that both methods are able to distinguish over 99% of the tested strains using eight SSR markers. The capillary sequencer was able to distinguish more strains while being faster; however, the accuracy of agarose gel electrophoresis was good enough to achieve almost the same results.

Luca Nerva and Walter Chitarra (eds.), *Botrytis cinerea*, Methods in Molecular Biology, vol. 3076,
https://doi.org/10.1007/978-1-0716-5551-1_9,

2 Materials

2.1 DNA Extraction

1. RED Extract Plant PCR kit (Merck KGaA, Darmstadt, Germany). For the contents of the kit, see [9].
2. Sterile forceps.
3. Microcentrifuge tubes, 1.5 mL.
4. Heat block or incubator.
5. Milli-Q water or otherwise purified water.

2.2 SSR-PCR and Agarose Gel Electrophoresis

1. Microcentrifuge tubes, 0.5 mL (*see* **Note 1**).
2. Make sure the primers used are specific for the tested species if necessary. Primer pairs should be highly conserved.
3. Preparation of primer mix: Evaluate all primers used for produced fragment sizes and temperature optimum (*see* **Notes 2 and 3**) and prepare the primer mixes for a final concentration of 10 pmol/μL each using Milli-Q water (*see* **Note 4**).
4. PCR Master Mix: Combine 10 μL RED Extract (Sigma-Aldrich, St. Louis, MO, USA), 4 μL Milli-Q water, and 1 μL each forward/reverse primer for each sample and each different primer mix. Make sure to include 10% overage for every master mix.
5. Agarose gel preparation: For the analysis with agarose gel electrophoresis, prepare a 3.1% agarose gel. To prepare, add 3.1 g of agarose (AppliChem GmbH, Darmstadt, Germany) to 100 mL 1× TAE buffer (AppliChem GmbH, Darmstadt, Germany; *see* **Note 5**). Mix well and microwave at 600 W for 3–4 min until smooth (*see* **Note 6**). Let the mixture cool down to 50°C and add 10 μL GelRed (GeneON, Ludwigshafen, Germany) and mix well (*see* **Note 7**). Pour the gel in an agarose gel chamber prepared with a comb. Let the gel cool down until solid.

2.3 Capillary Sequencer

1. 96-well plates.
2. Preparation of the primer mix: Evaluate all primers used for produced fragment sizes and temperature optimum (*see* **Note 8**) and prepare the primer mixes for a final concentration of 10 pmol/μL each using Milli-Q water. Mark the primer pairs with fluorescent labels (forward primer coupled with HEX, ROX, TAMRA, or FAM) so that primer pairs producing similar fragment sizes are coupled with a different label (*see* **Note 9**).
3. Use KAPA2G Fast Multiplex PCR Kit (KAPABIOSYSTENS, Wilmington, MA, USA) for the PCR.
4. Use Polymer POP7™ (Thermo Fisher Scientific, Waltham, MA, USA) for the sequencer.

3 Methods

3.1 DNA Extraction

1. The strains and samples were extracted using the RED Extract Plant PCR kit from Merck KGaA, Darmstadt, Germany.
2. All steps are carried out at room temperature unless otherwise noted.
3. Harvest mycelium from agar plates and from grape berries using sterile forceps. If possible, do not use more than 100 mg of starting material to ensure a good quality of DNA extraction (*see* **Note 10**).
4. Add 100 µL of extraction solution to the prepared sample and mix by vortexing.
5. Incubate in a heat block at 95°C for 10 min.
6. Subsequently, add 100 µL of dilution solution to the sample and mix by vortexing.
7. Centrifuge at 13,000 × *g* for 10 min and keep the soluble phase. Store in the freezer at −20°C for later use.

3.2 Agarose Gel Electrophoresis

1. Prepare the PCR run by adding 4 µL DNA sample to 16 µL of the prepared master mix for every tested primer mix in a microcentrifuge tube for a total reaction volume of 20 µL. Quickly centrifuge to collect the contents at the bottom of the tube (*see* **Note 11**).
2. Perform PCR as follows: Denaturation at 94°C for 30 s, followed by 36 cycles of: a specific primer-dependent temperature (here 50°C or 60°C) for 30 s, elongation at 72°C for 30 s, followed by an initial denaturation at 94°C for 3 min and another elongation at 72°C for 10 min.
3. Store the obtained DNA at −20°C in the freezer or at 4°C in the fridge for later use.
4. Load 5–10 µL of every PCR probe onto the prepared agarose gel (*see* **Note 12**).
5. Load 2 µL DNA ladder (Thermo Fisher Scientific, Waltham, MA, USA; *see* **Note 13**).
6. Run the gel at 90 V for 2 h and 45 min (*see* **Note 14**).
7. Capture the resulting gel using a UV camera lens.
8. Analyze the gel images (Fig. 1) using an image software tool such as ImageJ (version 1.51) [10] (*see* **Note 15**).

3.3 SSR PCR: Capillary Sequencer

1. Prepare the PCR run by adding 4 µL DNA sample to 16 µL of the prepared master mix for every tested primer (*see* **Note 15**) for a total of 20 µL reaction volume.
2. Mix in a 96-well plate.

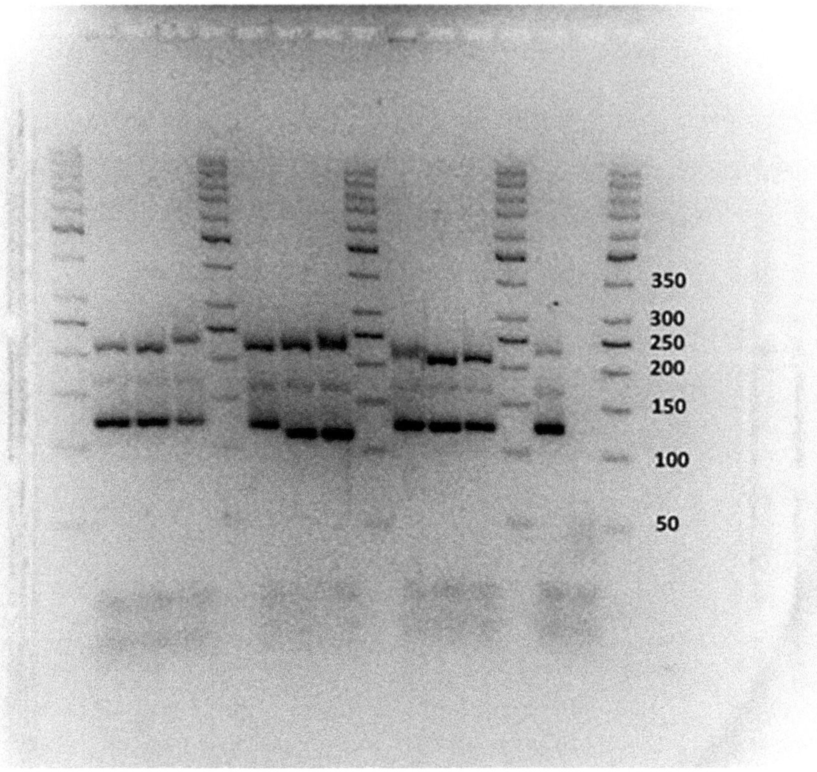

Fig. 1 Resulting gel image of an SSR-PCR analysis. 3.1% agarose gel, 90V run for 2 h 45 min

3. Perform PCR: the PCR program consisted of 30 cycles of denaturation, annealing, and elongation, with initial denaturation at 95°C for 3 min, denaturation at 95°C for 15 s, primer annealing at 60°C for 30 s, elongation at 72°C for 30 or 50 s, and final elongation at 72°C for 3 min.
4. Analyze the length of the fragments using a 3130xl Genetic Analyzer (Applied Biosystems, Darmstadt, Germany) and the corresponding GeneMapper 4.0. software.
5. Store the PCR samples at 4°C in the fridge or in the freezer (−20°C) for later use.

4 Notes

1. Using smaller microcentrifuge tubes ensures better incorporation and handling of the mixture.
2. It is important to test the primers for their optimal annealing temperature and specificity. Primer pairs can be checked for

specificity using available sites such as Ensembl. The annealing temperature can be checked via several online tools such as ThermoFisher Tm Calculator [11] or NEB Tm Calculator [12].

3. The primer mixes for agarose gel electrophoresis need to be checked for their potential resulting fragment sizes, annealing temperature, inhibition of one another, and cross-amplification. Checking these parameters via pretests ensures an optimal setup for the experiment and can reduce the amount of PCRs and gels.
4. Typical primer concentrations are 0.4 μM each. Adjusting the primer concentration depending on the system might be necessary. We found 1 μL of 10 μM for every primer best suitable, resulting in a final concentration of 0.5 μM (20 μL reaction volume).
5. We find 3.1% agarose gel concentration optimal for the obtained fragment sizes. However, this may be adjusted according to the expected fragment sizes if desired.
6. While microwaving the agarose-TAE mixture, boiling delay can occur, especially when using 3.1% agarose. Make sure to periodically shake the mixture during the microwave process. The mixture should be clear and free of any smears before pouring.
7. Temperatures above 50°C will destroy the GelRED used for staining the gel. Letting the mixture cool down too much will cause solidification and uneven dissolution of GelRED.
8. The capillary sequencer primers should be designed to work at the same annealing temperature. This will reduce the process time. Primer pairs with overlapping fragment sizes should be coupled with different fluorescent labels.
9. If collecting from agar plates, make sure to transfer no agar while transferring the mycelium. This can potentially inhibit the DNA extraction. Using more than 100 mg of mycelium can also potentially inhibit the extraction.
10. Collecting the contents of the PCR at the bottom of the microcentrifuge tubes ensures a reproducible and even reaction during PCR.
11. The PCR Kit comes with implemented loading dye for the gel electrophoresis. Measuring DNA content before loading on the gel may therefore give inaccurate results. We found 5–10 μL of DNA sample to be perfect for good-quality results in the gel.
12. We used 2 μL of DNA ladder five times per gel per run to ensure even separation of the fragments. The amount of DNA ladder can be adjusted to 0.5 μL or 1 μL if necessary.

13. We find 2 h and 45 min runtime at 90 V optimal for the run. This can be adjusted depending on the expected fragment sizes.
14. ImageJ was used to obtain the fragment sizes using the DNA ladder from each gel as a reference. The closest left and right DNA ladder for each sample per gel was used as a linear regression curve to even out uneven running of the gel.
15. If using multiple fluorescent dyes, multiple master mixes need to be assembled before adding the DNA samples.

5 Conclusion

This chapter provides two different methods for assessing and differentiating strains using SSR markers. The methods vary in complexity and cost, but both provided accurate results and were able to assign "fingerprints" to every tested strain.

Acknowledgments

This IGF Project (01IF21630N) of the FEI is supported within the program for promoting the Industrial Collective Research (IGF) of the Federal Ministry of Economic Affairs and Climate Action (BMWK), based on a resolution of the German Parliament. We would also like to thank Florian Schwander and Magrit Daum for help with the analysis on the capillary sequencer.

References

1. Fournier E et al (2002) Characterization of nine polymorphic microsatellite loci in the fungus *Botrytis cinerea* (Ascomycota). Mol Ecol Notes 2:253–255. https://doi.org/10.1046/j.1471-8286.2002.00207.x
2. Backmann L et al (2024) Molecular biological methods to assess different *Botrytis cinerea* strains on grapes. Microbiol Res 15:567–581. https://doi.org/10.3390/microbiolres15020037
3. Ibarra-Torres P et al (2015) Inter- and intraspecific differentiation of *Capsicum annuum* and *Capsicum pubescens* using ISSR and SSR markers. Sci Hortic 181:137–146. https://doi.org/10.1016/j.scienta.2014.10.054
4. Javier Gallego F et al (2005) Comparison of RAPDs, AFLPs and SSR markers for the genetic analysis of yeast strains of *Saccharomyces cerevisiae*. Food Microbiol 22:561–568. https://doi.org/10.1016/j.fm.2004.11.019
5. Luckey JA et al (1990) High speed DNA sequencing by capillary electrophoresis. Nucleic Acids Res 18:4417–4421. https://doi.org/10.1093/nar/18.15.4417
6. Vemireddy LR, Archak S, Nagaraju J (2007) Capillary electrophoresis is essential for microsatellite marker based detection and quantification of adulteration of basmati rice (*Oryza sativa*). JAgric Food Chem 55:8112–8117. https://doi.org/10.1021/jf0714517
7. Busch U, Nitschko H (1999) Methods for the differentiation of microorganisms. J Chromatogr B 722:263–278. https://doi.org/10.1016/s0378-4347(98)00369-7
8. Fawley WN et al (2015) Development and validation of an internationally standardized, high-resolution capillary gel-based electrophoresis PCR-ribotyping protocol for *Clostridium difficile*. PLoS One 10:1–14. https://doi.org/10.1371/journal.pone.0118150

9. Merck (2024) REDExtract-N-AMP Plant PCR Kit. https://www.sigmaaldrich.com/DE/de/product/sigma/xnap. Accessed 15 Nov 2024.
10. Schneider CA, Rasband WS, Eliceiri KW (2012) NIH Image to ImageJ: 25 years of image analysis. Nat Methods 9:671–675. https://doi.org/10.1038/nmeth.2089
11. Thermo Fisher Scientific (2024) Thermo Scientific web tools. https://www.thermofisher.com/de/de/home/brands/thermo-scientific/molecular-biology/molecular-biology-learning-center/molecular-biology-resource-library/thermo-scientific-web-tools.html. Accessed 15 Nov 2024
12. New England Biolabs (2024) Tm calculator. https://tmcalculator.neb.com/#!/main. Accessed 15 Nov 2024.

Chapter 10

Genome Assembly and Annotation of a Plant Pathogen Using the Example of *Botrytis cinerea*

Xiaoqian Shi-Kunne and Jan A. L. van Kan

Abstract

The availability of a high-quality genome assembly facilitates the analysis of fungal genomes. This chapter outlines the tools and steps involved in genome sequence assembly and annotation of a plant pathogen, *Botrytis cinerea*. We describe the use of Illumina short-read and Oxford Nanopore long-read sequencing data to assemble the *B. cinerea* genome. The steps include the pre-processing of sequencing data, genome assembly using Flye, scaffolding with NtLink, and polishing with Racon, Medaka, and NextPolish. The quality of the final assembly is evaluated using BUSCO, which serves as a benchmark for the completeness of a genome. We also provide details on the identification and masking of repetitive elements using the EarlGrey pipeline, as well as the gene prediction and annotation process with Funannotate. The methodologies and insights described can be applied to genome research in other fungal species.

Keywords Long-read sequencing, Genome assembly and polishing, Functional genome annotation, Repetitive element analysis, Bioinformatics pipelines

1 Introduction

As sequencing technologies have advanced, genomic studies have been boosted in many species, including fungi. The availability of a high-quality genome assembly facilitates the analysis of fungal genomes [1]. In many fungal plant pathogens, pathogenicity-related genes often reside in complex genomic areas with high repeat density, facilitating rapid evolution of these regions [2]. By having a complete genome assembly, we can not only discover pathogenicity-related genes but also uncover their evolutionary history. Genome assembly and annotation are fundamental steps in genome analyses that provide key insights into pathogenic fungi, such as *Botrytis cinerea* [3]. *Botrytis cinerea* is a notorious plant pathogen with the ability to infect a wide range of hosts. In this chapter, we will outline the steps and tools involved in genome assembly using Illumina and Nanopore reads, as well as the subsequent quality evaluation and annotation processes. While the focus is on the genome of *Botrytis*

Luca Nerva and Walter Chitarra (eds.), *Botrytis cinerea*, Methods in Molecular Biology, vol. 3076,
https://doi.org/10.1007/978-1-0716-5551-1_10,

cinerea, the methodologies and insights presented can be applied to genome research in other fungi as well.

2 Materials

Please note that all steps are run under the Linux system, and most required tools listed below can be installed via Conda (Table 1). Conda is an open-source environment management system. Miniconda is a minimal installer for Conda.

2.1 Software Tools and Pipelines

2.2 Sequencing Technologies

B. cinerea DNA samples for sequencing.

2.1.1 Biological Samples

Short-read (Illumina) and long-read (Oxford Nanopore) platforms (Table 2).

Table 1
Software tools and pipelines used in this analysis

Software	Purpose	Link
NanoPlot [4]	Nanopore read quality check	https://github.com/wdecoster/NanoPlot
Porechop	Adapter trimming	https://github.com/rrwick/Porechop
Chopper [4]	Read filtering	https://github.com/wdecoster/chopper
Flye [5]	Genome assembly	https://github.com/mikolmogorov/Flye
Racon	Assembly polishing	https://github.com/isovic/racon
Medaka	Assembly polishing	https://github.com/nanoporetech/medaka
NextPolish	Assembly polishing	https://github.com/Nextomics/NextPolish
Funannotate	Genome annotation	https://github.com/nextgenusfs/funannotate
QUAST	Assembly evaluation	https://github.com/ablab/quast
Tapestry	Assembly visualization	https://github.com/johnomics/tapestry
BBMap/BBtools	Bioinformatic tools	https://github.com/BioInfoTools/BBMap
BUSCO [6,7]	Assessment of assembly completeness	https://busco.ezlab.org/
EarlGrey [8]	Transposable element (TE) annotation	https://github.com/TobyBaril/EarlGrey?tab=readme-ov-file
MUMmer [9]	Genome aligner	https://mummer.sourceforge.net/

Table 2
Type of genomic data and file names used in this analysis

Type of reads	File name	Notes
Nanopore long reads	M3a.nano.fastq.gz	".gz" means the file is zipped
Nanopore long reads	M3a.nano.trimmed.old.fa	These reads do not have quality scores
Illumina DNA short reads	M3a_1.fastq	Forward read pair
Illumina DNA short reads	M3a_2.fastq	Reverse read pair
Illumina RNA short reads	M3a_RNA_1.fastq	Forward read pair
Illumina RNA short reads	M3a_RNA_2.fastq	Reverse read pair
Genome assembly of strain B05.10 [3]	Botrytis_cinerea.ASM83294v1.dna.fa	

3 Methods

3.1 Genome Assembly

After downloading and installing all required tools, all commands will be subsequently run in your terminal window as shown in the code boxes below. "#" is used in the code chunks below to denote a non-executable comment. The codes beginning without "#" can be copied, pasted, and run in your terminal.

Before starting the genome assembly process, it is crucial to evaluate the quality and characteristics of the Nanopore long-read sequencing data. This can be done using tools such as NanoPlot, which provides comprehensive summaries of the Oxford Nanopore sequencing data. Evaluating these statistics can help identify potential issues with the Nanopore data, such as low coverage, short read lengths, or poor quality. This information can then be used to make informed decisions about the assembly strategy, such as whether to supplement the Nanopore data with Illumina short-read sequencing to improve the final genome assembly.

- Checking long-read sequencing data

```
NanoPlot -t 4 --fastq M3a.nano.fastq.gz --plots dot
```

In the output file "NanoPlot-report.html," you will find comprehensive summaries of read statistics, including mean read length, total number of reads, total number of bases, etc. To calculate the coverage of the reads, you can use the total bases and the estimated genome size. In this example, we have a total base count of 327,386,3929 and a genome size of 42 Mb. Thus, the coverage can be calculated as follows:

$$\text{Coverage} = \text{Totalbases}/\text{Genomesize} \\ = 3,273,863,929/42,000,000 \\ \approx 77.95$$

This indicates a coverage of nearly 78×. Additionally, you will find figures illustrating the read length and quality, as shown below (Fig. 1). Usually, a Nanopore sequencing coverage of 60× to 100× is recommended to achieve a high-quality, contiguous assembly. This increased coverage helps resolve the repetitive regions and provides sufficient long-range information to bridge gaps and produce a more complete genome assembly.

Sequencing platforms typically add adaptor sequences to the DNA fragments during library preparation. These adaptor sequences need to be removed before the reads can be used for downstream analysis, as their presence will interfere with the assembly process. Sequencing data can also contain low-quality reads or regions, which negatively impact the process. Therefore, the next steps involve two critical processes: adaptor trimming and quality filtering.

- Pre-processing of sequencing data: adaptor trimming. We remove the adaptors present in the reads to ensure that only the actual sequence data is retained.

```
porechop -i M3a.nano.fastq.gz -o M3a.nano.trimmed.
fastq
```

- Pre-processing of sequencing quality: filtering. We filter out reads that have a quality score below 10. Reads with a quality score below 10 have a relatively high error rate, meaning there is a 1 in 10 chance that the base call is incorrect. The specific threshold of 10 is commonly used as it represents a reasonable balance between retaining a sufficient number of reads while excluding the most error-prone ones. Some studies may use a slightly different threshold, but 10 is a widely accepted and recommended value for Nanopore read quality filtering. The NanoPlot results (see Fig. 1) indicate that only a very small portion of reads have a quality score below 10. As a result, more than 95% of the reads will be retained in this analysis, ensuring high-quality data for further processing.

```
chopper -q 10 -l 500 -i M3a.nano.trimmed.fastq >
M3a.nano.trimmed.filtered.fastq
```

During the genome assembly process, it is common to incorporate read data from multiple sequencing experiments to improve the overall quality and contiguity of the final assembly. In this case, we are adding another set of Nanopore reads (21× coverage) to the assembly. However, this additional dataset does not include quality scores, which means we are unable to apply quality filtering in this case. This set of reads has a mean length of 7.7 kb and an N50 read length of 13 kb, which is

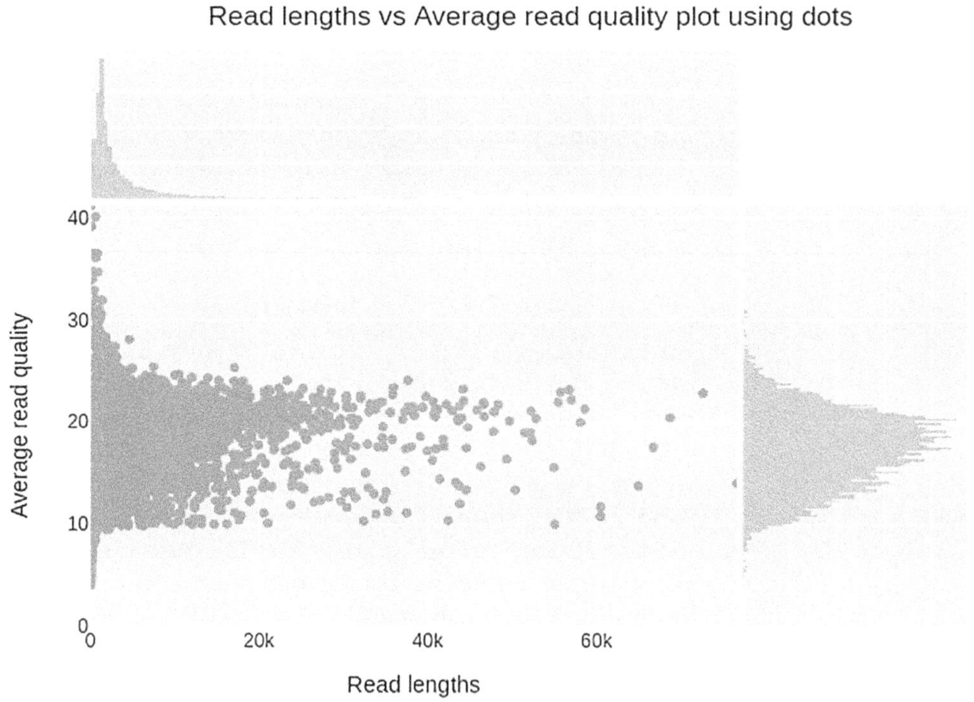

Fig. 1 Read lengths versus average read quality plot from NanoPlot. The scatter plot shows the relationship between read lengths and average read quality for a sequencing dataset. Each point represents a single read, with the *x*-axis showing the read length and the *y*-axis showing the average read quality. The plot provides insights into the quality and distribution of the sequencing reads

longer than the first set. Therefore, this additional set of reads can not only increase the total coverage but also add some even longer reads, which can facilitate the assembly of longer contigs.

- Assembly of *B.* cinerea genome using Flye.

```
flye --nano-raw M3a.nano.trimmed.filtered.fastq M3a.
nano.trimmed.old.fa --out-dir Flye_M3a --threads 20
```

- After running Flye, you will find the genome assembly in the main Flye output directory with the file name "assembly.fasta." You can optionally rename the genome file or create a symbolic link in another folder for the next step. Renaming is useful for keeping track of the processes completed in previous steps. Note: You do not have to perform all the commands listed below; you can instead only choose one or two depending on your specific aim.

```
#Rename the file
mv filtered_contigs.fasta M3a_flye.fa
#Copy the file to another folder/directory.
cp filtered_contigs.fasta /another_directory/M3a_
flye.fa
#Make a symbolic link in the folder of your next
step
```

```
ln -s /directory_of_the_assembly/filtered_contigs.
fasta M3a_flye.fa
```

- Although Flye provides a summary of the number of resulting contigs, you can check the number of contigs or obtain a detailed assembly evaluation by running QUAST

```
#quick check
grep ">"M3a_flye.fa |wc -l
#checking by quast
quast.py M3a_flye.fa -o quast_output
```

At the end of the assembly, we have obtained 37 contigs. However, we can further enhance the assembly by connecting some of the contigs through a process called scaffolding, using a tool called NtLink. NtLink employs minimizers to perform a lightweight mapping between the input target assembly and the supplied long reads. These long-read mapping data are then used as evidence to orient and order the output scaffolds.
We illustrate below the significance of read length in the scaffolding process by comparing two versions of the scaffolding results. In the first version, we used the file M3a.nano.trimmed.filtered.fastq, which contains reads with an average length of approximately 3 kb. In the second version, we also utilized the reads from M3a.nano.trimmed.old.fa, which have an average read length of 7.8 kb.

- Scaffolding only with "M3a.nano.trimmed.filtered.fastq". We run NtLink three times until there are no further changes in the number of scaffolds.

```
#First round
ntLink scaffold gap_fill overlap=True target= M3a_
flye.fa reads= M3a.nano.trimmed.filtered.fastq k=32
w=250
#Second round
ntLink scaffold gap_fill overlap=True target= M3a_
flye.fa .k32.w250.z1000.ntLink.scaffolds.gap_fill.fa
reads= M3a.nano.trimmed.filtered.fastq k=32 w=250
#Third round
ntLink scaffold gap_fill overlap=True target= M3a_
flye.fa .k32.w250.z1000.ntLink.scaffolds.gap_fill.fa.
k32.w250.z1000.ntLink.scaffolds.gap_fill.fa reads=
M3a.nano.trimmed.filtered.fastq k=32 w=250
#Change name of the file, you can also use "cp" to
keep the original file
mv M3a_flye.fa .k32.w250.z1000.ntLink.scaffolds.gap_
fill.fa. k32.w250.z1000.ntLink.scaffolds.gap_fill.fa.
k32.w250.z1000.ntLink.scaffolds.gap_fill.fa M3a_flye.
ntlink_V2.fa
#Count the number of contigs again
grep ">"M3a_flye.ntlink_V2.fa |wc -l
```

Here, we reduced the number of contigs from 37 to 25, which is pretty good. To validate the correctness of the scaffolds, we align the assembly to the genome of a well-assembled isolate, such as the reference B05.10.

- Run nucmer and mummer plot

```
#Run nucmer
nucmer -p B0510_vs_M3a Botrytis_cinerea.ASM83294v1.
dna.fa. M3a_flye.ntlink_V1.fa
#Run mummerplot
mummerplot -color B0510_vs_M3a_V2.delta
```

Overall, we achieved a fairly complete alignment of the M3a genome with the reference genome of B05.10 (Figs. 2 and 3). However, the largest contig, NtLink6, shows synteny with both chromosome 13 and chromosome 16 of the reference genome. This suggests that NtLink6 is likely a misassembly. Such misassembly, as

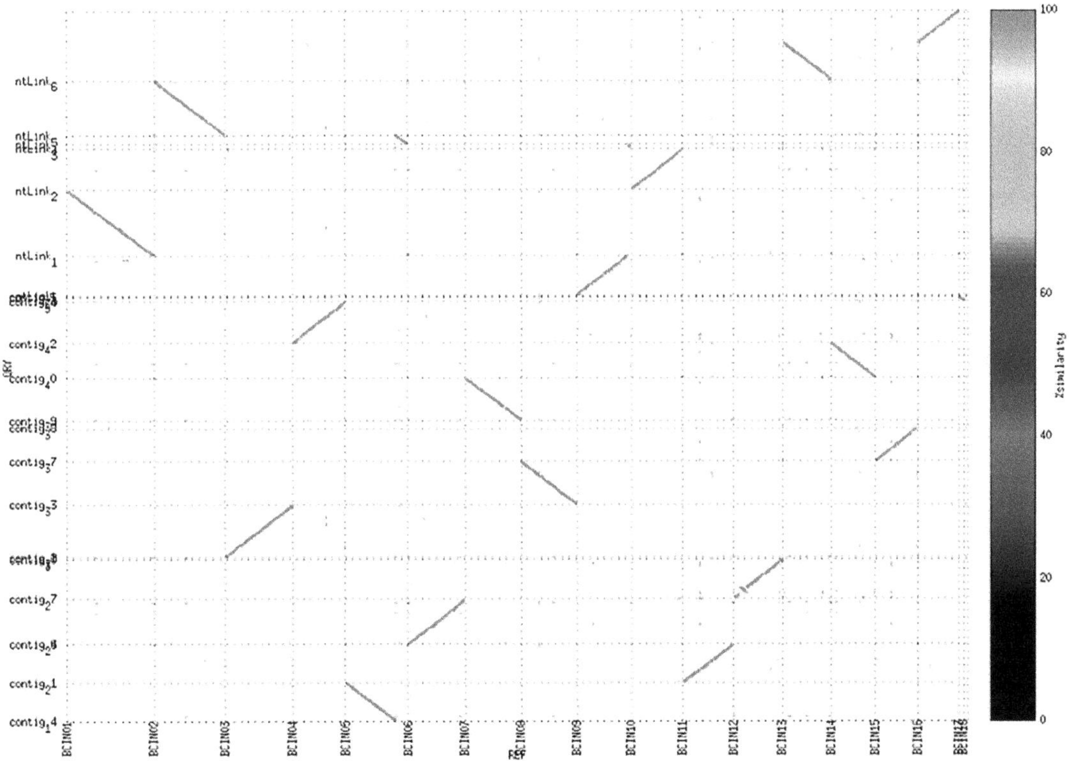

Fig. 2 Dotplot representation of the alignment between two genomic sequences. Each dot represents a match or alignment between the two genomes, with the *x*-axis (B05.10) and *y*-axis (M3a) corresponding to the coordinates of the respective genomes. The dotplot provides a visual overview of similarities and differences between the two genomes, allowing the identification of regions of high similarity (diagonal lines) as well as structural variations and rearrangements. Please note that the inverted diagonal lines that cross the whole grid from top left to bottom right result from the flipping of the orientation of contigs of M3a as compared to the reference chromosome (from B05.10), simply because the numberings start at opposite ends. These inverted lines are not indicative of true structural variation between the two genomic sequences

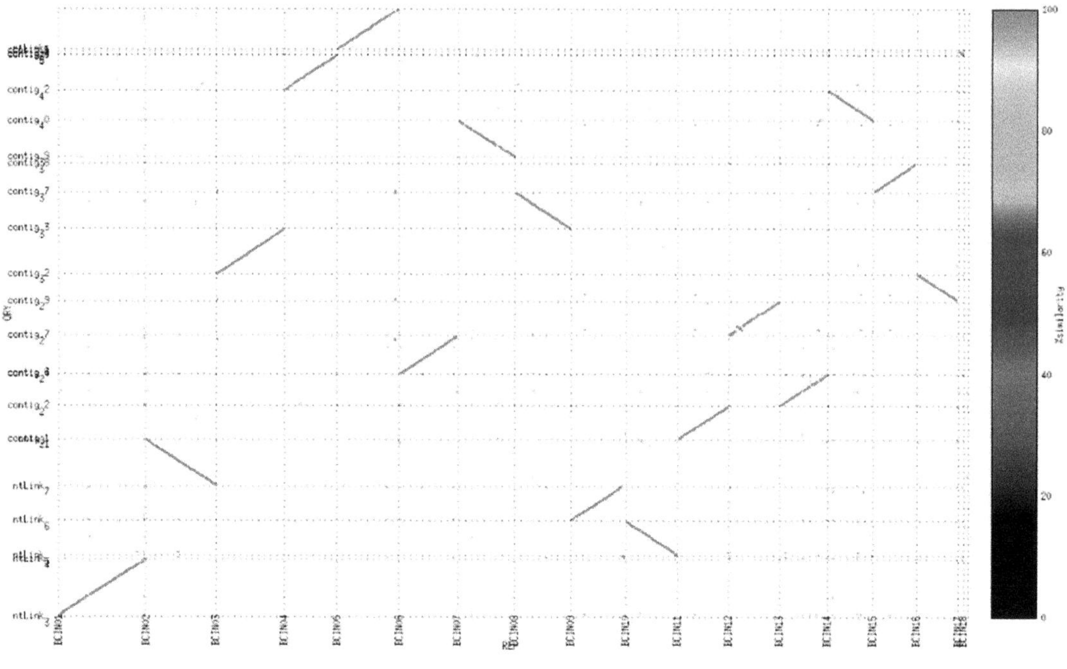

Fig. 3 Dotplot of genome alignment

observed with NtLink6, is largely attributed to the limitations of short reads in resolving complex genomic regions, especially duplicated repeats between chromosomes [10]. Short reads often struggle to accurately represent intricate structures, such as those found in telomeric regions, which include repetitive sequences and other challenging features. In this case, the inability of the short reads to span these complex areas likely led to incorrect alignments and the erroneous merging of distinct genomic segments, resulting in the observed synteny with both chromosome 13 and chromosome 16 of the reference genome. Using longer reads could potentially provide better resolution and help avoid misassembly by covering these complex regions more effectively. In the second version, we will use the longer reads from "M3a.nano.trimmed.old.fa." However, this set of reads lacks quality scores, which means we cannot apply filtering based on quality. As a result, we run NtLink with these longer reads only once. Following this step, we use the filtered reads from M3a.nano.trimmed.filtered.fastq to make further improvements to the assembly.

- Second version of scaffolding with "M3a.nano.trimmed.old.fa" and "M3a.nano.trimmed.filtered.fastq."

```
#First round
ntLink scaffold gap_fill overlap=True target= M3a_
flye.fa reads= M3a.nano.trimmed.old.fa k=32 w=250
#Second round
```

```
ntLink scaffold gap_fill overlap=True target= M3a_
flye.fa.k32.w250.z1000.ntLink.scaffolds.gap_fill.fa
reads= M3a.nano.trimmed.filtered.fastq k=32 w=250
#Third round
ntLink scaffold gap_fill overlap=True target= M3a_
flye.fa.k32.w250.z1000.ntLink.scaffolds.gap_fill.fa.
k32.w250.z1000.ntLink.scaffolds.gap_fill.fa reads=
M3a.nano.trimmed.filtered.fastq k=32 w=250
#Change name of the file, you can also use "cp" to
keep the original file
mv M3a_flye.fa.k32.w250.z1000.ntLink.scaffolds.gap_
fill.fa. k32.w250.z1000.ntLink.scaffolds.gap_fill.fa.
k32.w250.z1000.ntLink.scaffolds.gap_fill.fa M3a_flye.
ntlink_V2.fa
#Count the number of contigs again
grep ">"M3a_flye.ntlink_V2.fa |wc -l
```

Here, we reduced the number of contigs from 37 to 26. Again, we can validate the correctness of the scaffolds by aligning them to B05.10.

- Run nucmer and mummer plot

```
#Run nucmer
nucmer -p B0510_vs_M3a_V2 Botrytis_cinerea.
ASM83294v1.dna.fa. M3a_flye.ntlink_V2.fa
#Run mummerplot
mummerplot -color B0510_vs_M3a_V2.delta
```

In this version, we do not observe incorrectly joined contigs. Therefore, we proceeded with this version to the next step of the analysis. However, since Nanopore long reads contain a high number of errors (primarily INDELs), these errors will also be reflected in the genome assembly. In the next step, we corrected these errors in the assembly. According to a recent benchmark study on various tools, the best-performing polishing tool for long reads is Medaka, followed by Racon [11]. For short reads, the top tool is NextPolish. Therefore, we polished the genome assembly in the following order: Racon, Medaka, and then NextPolish. Before beginning the polishing process, we removed contigs that are either very small in size or represent the mitochondrial genome.

- We can begin by visualizing the assembly using Tapestry.

```
weave -a M3a_flye.ntlink_V2.fa -r M3a.nano.trimmed.
filtered.fastq -t TTAGGG -o M3a -c 10
```

Opening one of the resulting files, ending with ".html," produces an image as illustrated in Fig. 4. Based on this information, we removed some contigs: contig_71, contig_65, contig_53, contig_1, and contig_24. Contig_71 is very small in size and corresponds to the mitochondrial genome. Removing mitochondrial contigs is a common practice during the genome assembly process, as the mitochondrial genome is usually assembled separately from the nuclear

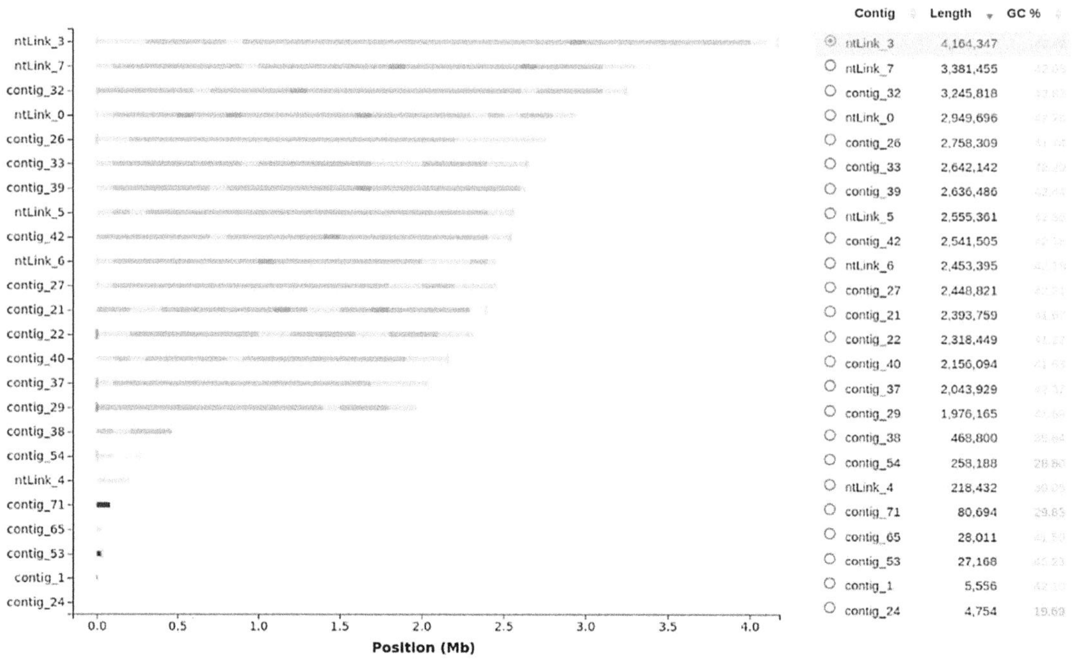

Fig. 4 Visualization of genome assembly contigs with read mapping density. The figure shows a representation of contigs of the genome assembly, with the density of the green color indicating the number of reads that have been mapped to each contig. Contigs are ordered by length. The red bars indicate telomeric repeats. This type of visualization provides an overview of the fragmentation and continuity of the assembled genome, as well as the distribution of sequencing reads across the contigs, which can be useful for assessing the quality and completeness of the genome assembly

genome. Contig_65, contig_53, contig_1, and contig_24 are also small, with lengths ranging from 5861 to 20,611 bp. Removing small contigs is a common step in the genome assembly process, as they often represent incomplete or low-quality genomic regions that may not contribute meaningful information. By removing small and mitochondrial contigs, the assembly is streamlined, focusing on the more complete and relevant genomic regions. This improves the overall quality and usability of the final genome assembly.

- Remove specific contigs.

```
filterbyname.sh in= M3a_flye.ntlink_V2.fa out= M3a_
flye.ntlink_V2_filtered.fa names=names.M3a.txt
```

- names.M3a.txt looks like:

```
contig_71
contig_65
contig_53
contig_24
```

Next, we use Racon for an initial correction of the assembly. According to the Medaka documentation, it is recommended to perform four rounds of correction with Racon before polishing with Medaka, as Medaka has been trained on Racon-polished assemblies. In this case, we run four rounds of Racon and only one round of Medaka corrections.

- Polish the genome assembly using Racon.

```
#1st round
minimap2 -t 20 -x map-ont flye.ntlink_V2_
filtered.fa M3a.nano.trimmed.filtered.fastq >mapIT01.
paf
racon -t 20 mapIT01.paf map-ont flye.ntlink_V2_
filtered.fa > mapIT01.fa
#2nd round
minimap2 -t 20 -x map-ont mapIT01.fa M3a.nano.
trimmed.filtered.fastq >mapIT02.paf
racon -t 20 mapIT02.paf map-ont mapIT01.fa >
mapIT02.fa
#3rd round
minimap2 -t 20 -x map-ont m Agarose gel
electrophoresis apIT02.fa M3a.nano.trimmed.filtered.
fastq >mapIT03.paf racon -t 20 mapIT03.paf map-ont
mapIT02.fa > mapIT03.fa
racon -t 20 mapIT03.paf map-ont mapIT02.fa >
mapIT03.fa
#4th round
minimap2 -t 20 -x map-ont mapIT03.fa M3a.nano.
trimmed.filtered.fastq >mapIT04.paf
racon -t 20 mapIT04.paf map-ont mapIT03.fa > M3a_
racon.fa
```

- Polish the genome assembly using Medaka. Note that the -m has to be modified based on your own read information. The model names follow this scheme: {pore}_{device}_{caller variant}_{caller version}.

```
medaka_consensus -d M3a_racon.fa -i M3a.nano.trimmed.
filtered.fastq -medaka_output -m r1041_e82_sup_g501
```

- Polish the genome assembly using NextPolish. Even after performing polishing with long reads, additional polishing with short reads remains essential. Long reads may not be fully corrected during long read polishing. Short reads can effectively identify and correct these errors due to their higher accuracy.

```
#Run NextPolish
nextPolish run.cfg
```

- Here, you can again change the file name. If you want to keep your original file, you can copy the file instead

```
#Change the file name
mv genome.nextpolish.fa M3a_polished.fa
# Or copy the file instead
cp genome.nextpolish.fa M3a_polished.fa
```

You can run a tool called BUSCO to assess the quality of your assembly. BUSCO determines whether well-established single-copy orthologs in specific lineages are correctly assembled or fragmented in contigs of a genome assembly. For instance, the BUSCO set for Ascomycota in odb10 contains a total of 1440 single-copy orthologs. This set is designed to assess the completeness of assemblies and annotations specifically for Ascomycetes. In a more contiguous genome assembly, complete BUSCO values are higher. BUSCO score above 95% is considered highly complete, with most of the expected single-copy orthologs present.

- Assessment of assembly quality using BUSCO

```
busco -i M3a_polished.fa -c 4 -o botrytis_cinerea
-m genome -l ascomycota _odb10
```

Here, we obtained BUSCO score of 99%, indicating a high-quality genome assembly with respect to the completeness of the expected gene content.

3.2 Genome Annotation

Funannotate is a pipeline specifically designed for genome annotation of fungal species. We used Funannotate to annotate the *B. cinerea* M3a genome.

- Prepare the genome by ranking and renaming the contigs by their sizes.

```
#Sort and rename all contigs
funannotate clean -i M3a_polished.fa -o M3a.final.
fasta
```

- Repeat prediction using EarlGrey

```
earlGrey -g M3a.final.fasta -s Botrytis_cinerea -o./
earlGreyOutputs
```

- For the gene prediction step, we use the masked genome file from the earlier EarlGrey process as the input file.

```
#Run Funannotate training to align RNA-seq data,
run Trinity, and then run PASA
funannotate train -i earlGreyOutputs /
botrytisCinerea_RepeatMasker_Against_Custom_
Library/ M3a.final.fasta.prep.masked -o fun_M3a
--left M3a_RNA_1.fastq --right M3a_RNA_2.fastq
--stranded FR --jaccard_clip --species "Botrytis
cinerea" --strain M3a --cpu 20
#Run Funannotate predict to run through the gene
prediction pipeline
funannotate predict -i M3a.final.fasta.masked.fa -o
fun_M3a --species " Botrytis cinerea " --strain M3a
--cpu 20
#Run Funannotate update to add UTR data to the
predictions and fix gene models that are in
disagreement with the RNA-seq data
funannotate update -i fun_M3a --cpus 20
```

In the end, Funannotate predicted about 11,400 protein-coding genes. The reference genome of B05.10 contains about 11,700 protein-coding genes. This difference is not surprising, as the annotation of B05.10 has been manually curated, allowing for more precise gene identification. Automated tools may miss or incorrectly annotate genes, especially in complex regions. Moreover, genes have complex features, such as alternative splicing, overlapping genes, and non-coding regions, that automated methods are often unable to accurately resolve. Although the detailed steps of manual annotation are not included here, it is highly recommended to use genome annotation editor tools (e.g., Apollo) to ensure high accuracy and provide valuable biological insights. Manual annotation often complements automated methods, leading to a more comprehensive understanding of genomic data.

References

1. Thomma BPHJ, Seidl MF, Shi-Kunne X, Cook DE, Bolton MD, van Kan JAL, Faino L (2016) Mind the gap; seven reasons to close fragmented genome assemblies. Fungal Genet Biol 90:24–30
2. Raffaele S, Kamoun S (2012) Genome evolution in filamentous plant pathogens: why bigger can be better. Nat Rev Microbiol 10(6): 417–430
3. Van Kan JAL, Stassen JHM, Mosbach A, Van Der Lee TAJ, Faino L, Farmer AD, Papasotiriou DG, Zhou S, Seidl MF, Cottam E, Edel D, Hahn M, Schwartz DC, Dietrich RA, Widdison S, Scalliet G (2017) A gapless genome sequence of the fungus *Botrytis cinerea*. Mol Plant Pathol 18(1):75–89
4. De Coster W, Rademakers R (2023) NanoPack2: population-scale evaluation of long-read sequencing data. Bioinformatics 39(5):btad311
5. Kolmogorov M, Bickhart DM, Behsaz B, Gurevich A, Rayko M, Shin SB, Kuhn K, Yuan J, Polevikov E, Smith TPL, Pevzner PA (2020) metaFlye: scalable long-read metagenome assembly using repeat graphs. Nat Methods 17(11):1103–1110
6. Simão FA, Waterhouse RM, Ioannidis P, Kriventseva EV, Zdobnov EM (2015) BUSCO: assessing genome assembly and annotation completeness with single-copy orthologs. Bioinformatics 31(19):3210–3212
7. Manni M, Berkeley MR, Seppey M, Simão FA, Zdobnov EM (2021) BUSCO update: novel and streamlined workflows along with broader and deeper phylogenetic coverage for scoring of eukaryotic, prokaryotic, and viral genomes. Mol Biol Evol 38(10):4647–4654
8. Baril T, Galbraith J, Hayward A (2024) Earl Grey: a fully automated user-friendly transposable element annotation and analysis pipeline. Mol Biol Evol 41(4):msae068
9. Kurtz S, Phillippy A, Delcher AL, Smoot M, Shumway M, Antonescu C, Salzberg SL (2004) Versatile and open software for comparing large genomes. Genome Biol 5(2):R12
10. Klug K, Zhu P, Pattar P, Mueller T, Safari N, Sommer F, Valero-Jiménez CA, van Kan JAL, Huettel B, Stueber K, Scheuring D, Schroda M, Hahn M (2024) Genome comparisons between *Botrytis fabae* and the closely related gray mold fungus *Botrytis cinerea* reveal possible explanations for their contrasting host ranges. J Fungi 10(3)
11. Luan T, Commichaux S, Hoffmann M, Jayeola V, Jang JH, Pop M, Rand H, Luo Y (2024) Benchmarking short and long read polishing tools for nanopore assemblies: achieving near-perfect genomes for outbreak isolates. BMC Genomics 25(1):679

Check for updates

Chapter 11

Identification of Mycoviral Infections

Ana Ruiz-Padilla, Julio L. Rodríguez-Romero, Marco Chiapello and María A. Ayllón

Abstract

High-throughput sequencing of total RNA has permitted the detection of novel mycoviruses in fungi with different types of genomes, including mycoviruses with double-stranded RNA, single-stranded positive- or negative-stranded RNA, or single-stranded DNA genomes. However, in silico detection of mycoviruses is not always sufficient to guarantee their presence in sequenced samples, especially in the case of the discovery of unique mycoviruses, and additional analyses are required to validate in vivo the data obtained by bioinformatics analysis. This chapter provides comprehensive protocols for the extraction of total RNA from the plant pathogenic fungus *Botrytis cinerea* for next-generation sequencing (NGS), outlines the bioinformatics pipeline designed and followed to detect mycoviruses in the sequenced samples, and details the detection in vivo and the complete molecular characterization of the mycoviruses identified in silico.

Keywords Next-generation sequencing (NGS), Mycoviruses, Virome, Viral metagenomics, Bioinformatic pipeline, Necrotrophic fungi, *Botrytis cinerea*

1 Introduction

Plant pathogenic fungi induce a variety of diseases in plants in the field, as well as in products in postharvest, causing substantial crop losses for the agricultural industry and becoming major threats to global food security. Damage includes reduced yields, loss of fresh and postharvest crops, and contamination by mycotoxins. *Botrytis* species, a group of widespread plant pathogenic necrotrophic fungi, cause gray mold or gray rot disease in many economically important crops [1]. *Botrytis cinerea* is considered the second most significant fungal plant pathogen worldwide [2]. The fungus is very destructive, causing enormous losses in more than 200 crops spread throughout the world, such as horticultural crops (tomato, cucumber, bean, pepper, etc.), grapevine, and strawberry plants, both in the field and in postharvest fruits. Different strategies have been used to control the fungus, such as cultural approaches, host resistance, and fungicides. However, control mainly depends on the application of a large

Luca Nerva and Walter Chitarra (eds.), *Botrytis cinerea*, Methods in Molecular Biology, vol. 3076,
https://doi.org/10.1007/978-1-0716-5551-1_11,

amount of fungicides that lead to rapid development of resistance in the fungus [3–5]. Overall, the constant threat of the emergence of resistant isolates, the growing public concern about crop chemical residues, and the secondary effects on human health have forced the search for safer alternatives to chemical control of *B. cinerea*. A safe alternative could be the use of biological or biotechnological control based on the use of mycoviruses.

Mycoviruses are viruses that infect fungi. They can have a positive influence on the infected fungal host, increasing its thermal tolerance [6] or changing its lifestyle by converting the fungus from a pathogen to a beneficial endophyte that enhances plant protection and production [7]. In some cases, they can also increase the virulence of their fungal hosts [8–10]. There are examples of mycoviruses that have no apparent impact on their fungal host; therefore, they can be potential candidates for the development of mycoviral infectious clones for reverse genetics in fungi [11,12]. Lastly, mycoviruses can also cause disease in infected fungi and decrease their virulence on infected plant hosts [13,14]. These mycovirus-mediated hypovirulent fungal strains can be potential biocontrol agents for use in biological control strategies against plant pathogenic fungi [15–18]. The use of viruses in control strategies for plant pathogens has been reviewed by Wagemans and coworkers [19]. To ascribe a fungal feature to the presence of a certain mycovirus, we need to determine the mycovirome of the fungus. Next-generation sequencing (NGS) of total RNA has allowed the fast discovery of novel and unique mycoviruses in many fungi and oomycetes [14,20–31]. Mycoviruses with a double-stranded RNA (dsRNA) or a positive-sense single-stranded RNA (ssRNA(+)) genome are very frequent, but there are also mycoviruses with a negative-sense single-stranded RNA (ssRNA(−)) genome, and only a few of them have a single-stranded DNA (ssDNA) genome [30,32–36]. In this chapter, we detail the complete procedure to obtain the mycovirome of *B. cinerea*, starting with the total RNA extraction that will be used for NGS, the completely detailed bioinformatics pipeline applied for the discovery of mycoviruses, and the procedures for the in vivo detection and characterization of these newly identified mycoviruses.

2 Materials

2.1 RNA Extraction

1. Potato dextrose agar (PDA) and potato dextrose broth (PDB)
2. Safe-lock tubes.
3. Glass beads (0.1 mm).
4. Bead beater (Qiagen TissueLyser) or vortex.
5. Spectrum Plant Total RNA Kit (Sigma-Aldrich).

6. Low EEO agarose.
7. Electrophoresis equipment.
8. Nanodrop.
9. 6× Gel Loading Dye.
10. RNA or DNA ladder.
11. DEPC water.
12. Miracloth.
13. Sterile tips.

2.2 Software

1. R software (version 4.3.2) https://www.r-project.org/. RStudio (Build 369) Desktop https://posit.co/ (optional), and R libraries: tidyverse_2.0.0; DT_0.31; fs_1.6.3; Biostrings_2.70.1; ggthemes_5.0.0; cowplot_1.1.3; network_1.18.2; ggraph_2.1.0; tidygraph_1.3.0.; BiocManager_1.30.22.
2. Trinity (version 2.8.4) [37].
3. BBTools (version 39.01–0) [38].
4. DIAMOND (version 0.9.22) [39].
5. Samtools (version 1.9) [40].
6. BWA (version 0.7.17) [41].
7. CAP3 (version https://github.com/brewsci/homebrew-bio/blob/master/Formula/cap3.rb) [42].
8. Tablet software (version 1.21.02.08) [43].
9. Basic Local Alignment Search Tool (BLAST) [44].
10. Primer3 web (Primer3 Input).
11. Primer-BLAST (Primer designing tool (nih.gov)).

2.3 Detection of Viruses by RT-qPCR

1. High-Capacity cDNA Reverse Transcription Kit (Applied Biosystems).
2. Specific primers for viral genomes.
3. Universal SYBR Green Master Mix.
4. LightCycler® 480 Multiwell Plate 384.
5. LightCycler® 480.
6. Centrifuge for multiwell plates.
7. Optionally, electrophoresis equipment, agarose, DNA ladder, and 6× gel loading dye.

2.4 Detection of Viruses by RT-PCR

1. SuperScript IV (Thermo Fisher Scientific).
2. Specific primers for viral genomes.
3. dNTPs (10 mM)

4. CloneAmp HiFi PCR (TaKaRa).
5. Thermal cycler.
6. Electrophoresis equipment.
7. Agarose.
8. 6× gel loading dye.
9. DNA ladder.

2.5 Determination of 5' and 3' terminal ends

1. SMARTer RACE Kit (Takara Bio) or RLM-RACE Kit.
2. Specific primers for viral genomes.
3. CloneAmp Taq Polymerase.
4. Thermal block.
5. Thermal cyclers.
6. Electrophoresis equipment.
7. Agarose.
8. 6× gel loading dye.
9. DNA ladder.

2.6 Combined PCRs

1. Kit for cDNA synthesis, including random primers, for example, High-Capacity cDNA Reverse Transcription Kit (Applied Biosystems).
2. Specific primers for viral genomes.
3. Thermal cycler.
4. Electrophoresis equipment.
5. Agarose.
6. 6× gel loading dye.
7. DNA ladders.

2.7 In Silico Search of Additional Segments in Multisegmented Genomes

1. Ubuntu desktop or Ubuntu operating system.
2. Microsoft Excel.

3 Methods

3.1 Total RNA Extraction for RNA-seq Analysis

Previously, isolates cultured on PDA are used to prepare liquid cultures in PDB. Briefly, a sterile tip is used to scratch mycelia and conidia from the PDA plate and transfer them to a PDB plate. Mycelia are collected after 3 days and dried with Miracloth, and total RNA is extracted with the Spectrum Plant Total RNA Kit (*see* **Note 1**).

1. Mix approximately 100 mg of dry mycelia (cultured in PDB) with 500 μL of Lysis Buffer and 5 μL of β-mercaptoethanol and disrupt using a Qiagen TissueLyser II.
2. Incubate the solutions at 56°C for 5 min and follow the manufacturer's instructions from this step onward.
3. Analyze 200 ng of extracted total RNA by electrophoresis (Fig. 1).

3.2 Bioinformatics Pipeline

The bioinformatics pipeline for in silico assembly and identification of mycoviruses, as devised by Chiapello and coworkers [45], encompasses seven key steps (Fig. 2):

a) Preprocessing reads.
b) Creating a custom viral database.
c) Conducting *de novo* assembly with the cleaned reads.

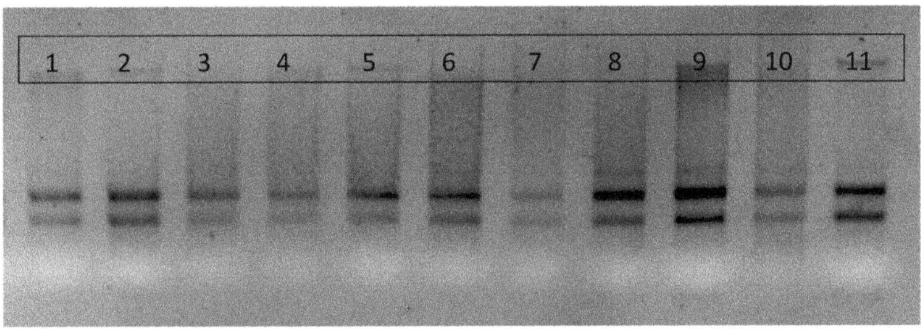

Fig. 1 Total RNA samples analyzed on a 1% agarose gel using 1× TAE (70 V, 45 min)

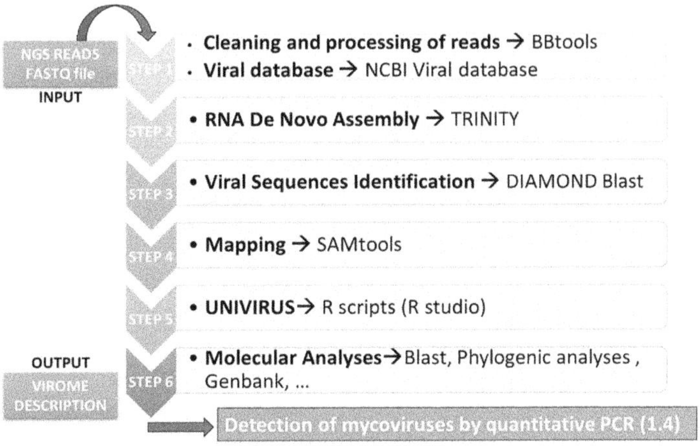

Fig. 2 Summary of the processing pipeline

d) Identifying viral sequences through a high-performance sequence aligner.

e) Mapping reads.

f) Selecting "Univirus."

g) Concluding with final molecular characterization.

3.2.1 Clean Reads and Prepare a Custom Viral Database

Following NGS sequencing, a FASTQ file containing paired-end reads will be generated for each sample. Clean reads will be processed using BBTools [38], where adapters, artifacts, short reads, and residual ribosomal sequences will be sequentially removed (Fig. 3). Ensure that the read file names adhere to the format "sampleID_1.fastq.gz" and "sampleID_2.fastq.gz" for paired ends, as these serve as input files for the pipeline.

For enhanced efficiency, the cleaning process can be automated through the execution of a single script. You can conveniently download these scripts from our *GitHub* repository.

Usage: $ bash clean_sample.sh sampleID

clean_sample.sh

```
#!/bin/sh

resources="./bbmap-39.01-0/resources/"
#Sketch
sendsketch.sh in=./sampleID/sampleID_*_1.fq.gz
out=./sampleID/sketch_1.txt reads=200000 fname=
sampleID_*_1.fq.gz minprob=0.2 samplerate=1.0 merge
```

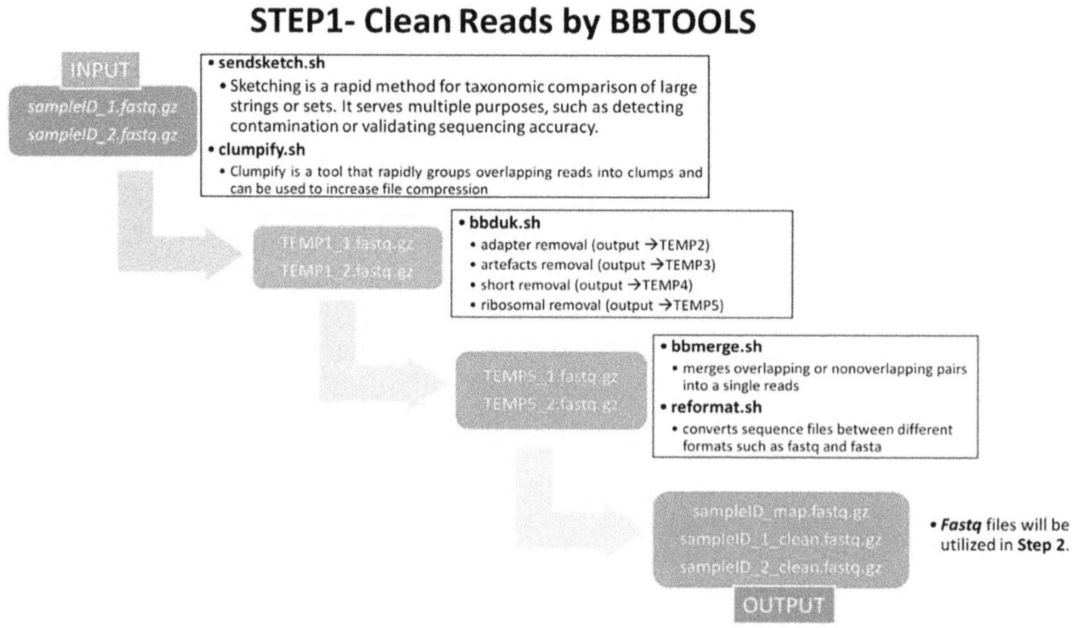

Fig. 3 Schematic representation of the use of BBTools to clean reads

```
printname0=f records=20 overwrite=true color=false
depth depth2 unique2 volume sortbyvolume contam2=
genus nt ow
sendsketch.sh in=./sampleID/sampleID_2.fastq.gz
out=./sampleID/sketch_2.txt reads=200000 fname=
sampleID.fastq.gz minprob=0.2 samplerate=1.0 merge
printname0=f records=20 overwrite=true color=false
depth depth2 unique2 volume sortbyvolume contam2=
genus nt ow
sendsketch.sh in=./sampleID/sampleID_1.fastq.gz
out=./sampleID/sketch_1.txt reads=200000 fname=
sampleID.fastq.gz minprob=0.2 samplerate=1.0 merge
printname0=f records=20 overwrite=true color=false
depth depth2 unique2 volume sortbyvolume contam2=
genus refseq append
sendsketch.sh in=./sampleID/sampleID_2.fastq.gz
out=./sampleID/sketch_2.txt reads=200000 fname=
sampleID.fastq.gz minprob=0.2 samplerate=1.0 merge
printname0=f records=20 overwrite=true color=false
depth depth2 unique2 volume sortbyvolume contam2=
genus refseq append
sendsketch.sh in=./sampleID/sampleID_1.fastq.gz
out=./sampleID/sketch_1.txt reads=200000 fname=
sampleID.fastq.gz minprob=0.2 samplerate=1.0 merge
printname0=f records=20 overwrite=true color=false
depth depth2 unique2 volume sortbyvolume contam2=
genus silva minkeycount=2 append
sendsketch.sh in=./sampleID/sampleID_2.fastq.gz
out=./sampleID/sketch_2.txt reads=200000 fname=
sampleID.fastq.gz minprob=0.2 samplerate=1.0 merge
printname0=f records=20 overwrite=true color=false
depth depth2 unique2 volume sortbyvolume contam2=
genus silva minkeycount=2 append
#Clumpify
clumpify.sh pigz=t unpigz=t zl=4 reorder in1=./
sampleID/sampleID_1.fastq.gz out1=./sampleID/
TEMP1_1.fastq.gz in2=./sampleID/sampleID_2.fastq.gz
out2=./sampleID/TEMP1_2.fastq.gz passes=1
#CLEAN Step1 - adapter removal
bbduk.sh -threads=8 ktrim=r ordered minlen=
49 minlenfraction=0.33 mink=11 tbo tpe rcomp=t
overwrite=true k=23 hdist=1 hdist2=1 ftm=5 pigz=t
unpigz=t zl=4 ow=true in1=./sampleID/TEMP1_1.fastq.
gz out1=./sampleID/TEMP2_1.fastq.gz in2=./sampleID/
TEMP1_2.fastq.gz out2=./sampleID/TEMP2_2.fastq.
gz rqc=hashmap outduk=./sampleID/ktrim_kmerStats1.
txt stats=./sampleID/ktrim_scaffoldStats1.txt loglog
ref=resources/adapters.fa
#CLEAN Step2 - artefacts removal
bbduk.sh -threads=8 maq=10,0 trimq=25 qtrim=
r ordered overwrite=true maxns=1 minlen=49
minlenfraction=0.33 k=25 hdist=1 pigz=t unpigz=
t zl=6 cf=t barcodefilter=crash ow=true in1=./
sampleID/TEMP2_1.fastq.gz out1=./sampleID/TEMP3_1.
```

```
fastq.gz in2=./sampleID/TEMP2_2.fastq.gz out2=./
sampleID/TEMP3_2.fastq.gz outm=./sampleID/synth1.
fq.gz rqc=hashmap outduk=./sampleID/kmerStats1.
txt stats=./sampleID/scaffoldStats1.txt loglog ref=
$resources/phix_adapters.fa.gz,$resources/lambda.
fa.gz,$resources/sequencing_artifacts.fa.gz
#CLEAN Step3 - short removal
bbduk.sh -threads=8 ordered overwrite=true k=
20 hdist=1 pigz=t unpigz=t zl=6 ow=true in1=./
sampleID/TEMP3_1.fastq.gz out1=./sampleID/TEMP4_1.
fastq.gz in2=./sampleID/TEMP3_2.fastq.gz out2=./
sampleID/TEMP4_2.fastq.gz outm=./sampleID/synth2.
fq.gz outduk=./sampleID/kmerStats2.txt stats=./
sampleID/scaffoldStats2.txt loglog ref=$resources/
short.fa
#CLEAN Step4 - ribosomal removal
bbduk.sh -threads=8 ordered k=31 ref=$resources/
ribokmers.fa.gz ow=true in1=./sampleID/TEMP4_1.
fastq.gz out1=./sampleID/TEMP5_1.fastq.gz in2=./
sampleID/TEMP4_2.fastq.gz out2=./sampleID/TEMP5_2.
fastq.gz outm=./sampleID/ribo.fq.gz outduk=./
sampleID/ribo_Stats1.txt stats=./sampleID/ribo_
Stats2.txt
#MERGE PAIRED-END
bbmerge.sh -threads=8 loose overwrite=true in1=./
sampleID/TEMP5_1.fastq.gz in2=./sampleID/TEMP5_2.
fastq.gz ihist=./sampleID/ihist_merge.txt outc=
./sampleID/cardinality.txt pigz=t unpigz=t zl=9
adapters=$resources/adapters.fa
reformat.sh in=./sampleID/TEMP5_#.fastq.gz out=./
sampleID/sampleID_map.fastq.gz
```

The outcomes will be stored in the designated *sampleID* folder. While this process is executable on a standard PC, leveraging a robust cluster would offer added convenience.

3.2.2 Prepare a Viral Database

Viral sequences will be identified by querying a custom or comprehensive viral database. To obtain the viral database in FASTA format, you may download the entire viral database by visiting the Index of /refseq/release/viral (nih.gov). Alternatively, you can use the following UNIX command line:

`$wget:` https://ftp.ncbi.nlm.nih.gov/refseq/release/viral/viral.1.protein.faa.gz"

For an alternative approach, consider utilizing EDirect. Entrez Direct (EDirect) streamlines access to NCBI's interconnected databases, encompassing viral sequences, through a Unix terminal window. Follow the steps below to install EDirect and download a protein viral database in FASTA format:

```
# Install EDirect
$ sh -c "$(curl -fsSL
)"
```

```
$ sh -c "$(wget -q
-O -)"
$ export PATH=${HOME}/edirect:${PATH}
# EDirect for viruses protein database
$ esearch -db protein -query "viruses[organism] NOT
srcdb pdb[prop]" | efetch -format fasta > ViralDB.
fasta
```

The resulting FASTA file, named ViralDB.fasta, can be employed for BLAST analysis. To utilize this FASTA file in DIAMOND software, you must construct a database by executing the following command:

```
$ diamond makedb --in ViralDB.fasta -d./viraldb
```

This will generate a database named viralDB. It is noteworthy that this process is compatible with a standard PC.

3.2.3 *RNA* De Novo *Assembly*

The results from the cleaning step (3.2.1) will serve as input for the subsequent *de novo* assembly. Output files from this step include "sampleID_1_clean.fastq.gz" and "sampleID_2_clean.fastq.gz." Trinity software [37] will be employed for the assembly process, necessitating substantial processing power (CPU). We strongly advise utilizing a robust cluster or opting for the Galaxy server (https://usegalaxy.org/) to accomplish this task efficiently.

Galaxy: The files "sampleID_1_clean.fastq.gz" and "sampleID_2_clean.fastq.gz" are uploaded to Galaxy.eu. Subsequently, Trinity is executed with standard parameters, a process that typically takes a few hours to complete. Once concluded, download the resulting FASTA file and store it in the "./results/sampleID" folder, ensuring it is renamed as "Trinity_sampleID.fasta."

Cluster or PC: A standard PC may struggle to run TRINITY effectively, as it demands high computing power. Therefore, we recommend utilizing a cluster for optimal performance. For local processing, adjust your CPU and Max_memory requirements to match your machine specifications (e.g., 120G), and execute Trinity with the following parameters:

#RNA Assembly by Trinity

```
$Trinity --max_memory 120G --seqType fq -left./
sampleID_1_clean.fastq.g -right./sampleID_1_2_
clean.fastq.gz --CPU 24 -output./Trinity_sampleID.
fasta --no_normalize_reads -- no_version_check
--jaccard_clip --full_cleanup
```

The output file from step 2 will be "Trinity_sampleID.fasta."

A representation of one of the transcript FASTA entries is provided below:

>TRINITY_DN1000_c115_g5_i1 len=247 path=[31015:0-148 23018:149-246]

AATCTTTTTTGGTATTGGCAGTACTGTGCTCTGGG-TAGTGATTAGGGCAAAAGAAGACAC

ACAATAAAGAACCAGGTGTTAGACGTCAGCAAGTCAA-GGCCTTGGTTCTCAGCAGACAGA

AGACAGCCCTTCTCAATCCTCATCCCTTCCCTGAACA-GACATGTCTTCTGCAAGCTTCTC

CAAGTCAGTTGTTCACAGGAACATCATCAGAATAAAT-TTGAAATTATGATTAGTATCTGA

TAAAGCA

The accession "TRINITY_DN1000_c115_g5_i1" decodes the Trinity read cluster "TRINITY_DN1000_c115," corresponding to gene "g5," and isoform "i1." In addition to these details, the FASTA header incorporates the nucleotide length and the path used for constructing these contigs.

This FASTA file encapsulates comprehensive *de novo* RNA scaffolds, meticulously assembled from reads. While the current assembly stands as a complete representation, an extended and potentially more intricate assembly can be achieved using Cap3 Contig software [42]. Cap3 Contig merges overlapping scaffold sequences from *Trinity_sampleID.fasta*, creating a set of new "Contigs" (Fig. 4). However, it is crucial to note that our analysis has identified potential artifacts generated by Cap3, which could impede virus analysis and detection in subsequent stages. Consequently, we advise against the use of Cap3 for this purpose.

3.2.4 Viral Sequences Identification

In this pivotal phase, the identification of viral sequences will be accomplished by comparing the custom viral database (as detailed in Sect. 3.2.1) against all scaffolds and contigs generated through *de novo* assembly (outlined in Sect. 3.2.3). For this purpose, we will leverage DIAMOND software [39], comparing ViralDB (as introduced in Sect. 3.2.2) with our *de novo*-assembled transcriptome file (Trinity_sampleID.fasta) (Fig. 5).

Execute DIAMOND BLASTX with the following parameters:

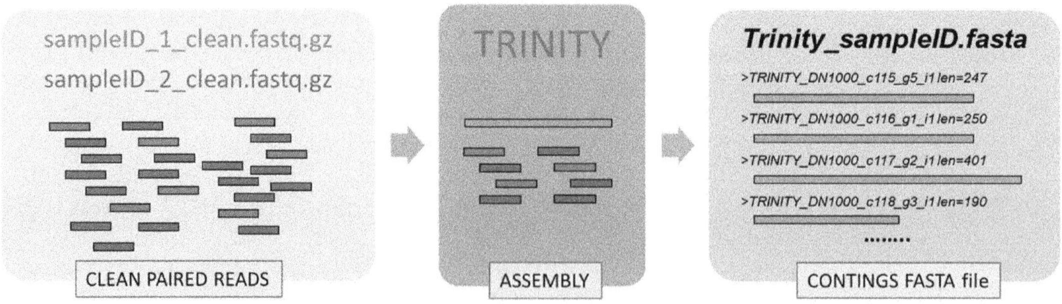

Fig. 4 Schematic representation of the use of *Cap3* Contig to merge overlapping scaffold sequences from Trinity_sampleID.fasta to create a set of new "Contigs"

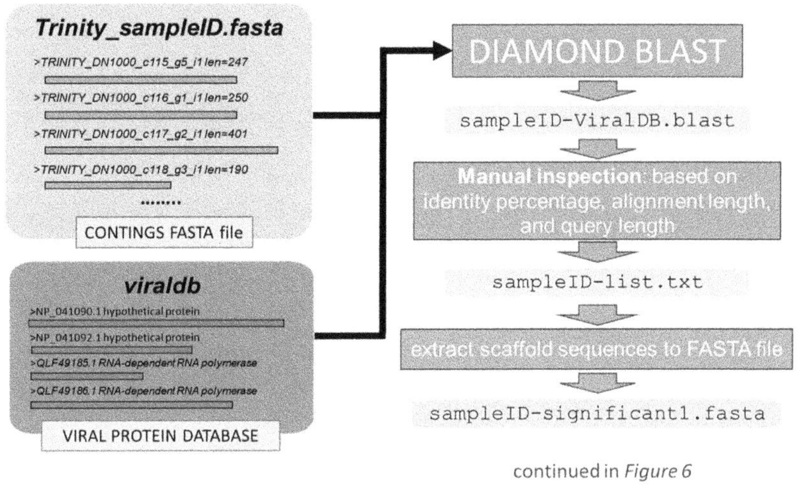

Fig. 5 Schematic representation of the use of DIAMOND software

```
#Diamond blast viraldb
$diamond blastx -p24 -d./viraldb -q./Trinity_
sampleID.fasta -o./sampleID-ViralDB.blast -f 0 -k 2
-e 0.00001 --unal 0 --more-sensitive
```

The resulting file, "sampleID-ViralDB.blast," can be opened in UNIX using commands such as "less" or "more." Subsequently, manual inspection of the results is conducted, and reliable viral sequences are selected based on identity percentage, alignment length, and query length.

Next, all scaffold names meeting our criteria are manually compiled into a text file, which is stored in the directory "./results/sampleID/sampleID-list.txt."

To extract the sequences, initiate the process by using "grep" to extract the "Query" names, and then utilize "cat" to consolidate these IDs into a unified list.

```
#Extract Contigs IDs list
$grep -e "Query=" $results/ sampleID-ViralDB.blast
>./results/sampleID/sampleID-list.txt
$cat./results/sampleID/sampleID-list.txt | sort |
uniq >./results/sampleID/sampleID-listuniq.txt
$cut -d " " -f 2./results/sampleID/sampleID-
listuniq.txt >./results/sampleID/sampleID-list.txt
```

With this compiled list, extract scaffold sequences from the original file "Trinity_sampleID.fasta." Employ the "awk" and "grep" commands to efficiently extract FASTA entries from the original Trinity file.

```
#Extract fasta from Trinity_sampleID.fasta
```

```
$awk'{if($0 ~ /^>/){print $0} else {printf $0}}'
Trinity_sampleID.fasta | perl - pe "s/>/\n>/g" >
Trinity_sampleIDoneline.fasta
$grep -A1 --no-group-separator -f./results/
sampleID/sampleID-list.txt Trinity_sampleIDoneline.
fasta > sampleID-significant1.fasta
```

The output file is *sampleID-significant1.fasta*

To distinguish between genuine viruses and integrated host sequences, a secondary BLAST analysis will be conducted. Obtain the completeNCBI nr database by directly downloading it from https://ftp.ncbi.nlm.nih.gov/blast/db/FASTA/nr.gz or by using the Unix command line: "wget https://ftp.ncbi.nlm.nih.gov/ blast/ db/FASTA/nr.gz -c"

#Download NCBI nr database and make a DIAMOND database

$wget https://ftp.ncbi.nlm.nih.gov/blast/db/FASTA/nr.gz -c

To utilize this FASTA file in DIAMOND software, you must construct a database by executing the following command:

```
$diamond makedb --in./nr.gz -d./nr
ase (
# Diamond blast using NCBI nr database
$ diamond blastx -d nr -q sampleID-significant1.
fasta -o sampleID-NCBInrblast.blast -f 0 -k 2 -e
0.00001 --unal 0 --more-sensitive
$ diamond blastx -d nr -q sampleID-significant1.
fasta -o sampleID-NCBInrtable.blast -f 6 qseqid
sseqid pident length qlen mismatch gapopen qstart
qend sstart send evalue bitscore -k 2 -e 0.00001
--unal 0 --more-sensitive
# Convert blast results into a table with headers
filename="sampleID-NCBInrblast.blast"
cut -d "[" -f 2 $filename | sed 's/]//g' > list1.txt
cut -d "[" -f 1 $filename | sed 's/>//g' > list2.txt
paste sampleID-NCBInrtable.blast list2.txt | paste
- list1.txt > NCBItable2.blast
echo "qseqid sseqid pident length qlen mismatch
gapopen qstart qend sstart send evalue bitscore
protein organism" > header
cat <(head -n 1 header) NCBItable2.blast >
sampleID-NCBInr.blast
```

The resulting output file is "sampleID-NCBInr.blast." Following this, a manual selection of scaffolds meeting our criteria becomes imperative. Save the resulting list in a text file located at "./results/ sampleID/sampleID-list2.txt." To extract sequences, commence by using "grep" to isolate query names and then utilize "cat" to amalgamate IDs into a list, as previously described. Leveraging this list, extract significant scaffold sequences from the file "sampleID-significant1.fasta" in FASTA format. Employ commands such as "awk" and "grep" to efficiently extract these sequences from the original

Trinity file, as elucidated earlier. While this process is executable on a standard PC, leveraging a robust cluster would be advantageous.

The resulting output file is "sampleID-significant2.fasta" (Fig. 6).

3.2.5 Mapping

The mapping of selected virus sequences with clean reads will be conducted using BWA, followed by transformation with SAMtools [40], and visualization through Tablet software [43]. Utilize the following shell script for this process:

```
# Mapping with bwa
$bwa index sampleID-significant2.fasta
$bwa mem -t 40 sampleID-significant2.fasta sampleID_
map.fastq.gz > sampleID-significant2.sam
$samtools view -bS sampleID-significant2.sam -o
sampleID-significant2.bam
$samtools sort -@ 40 sampleID-significant2.bam -o
sampleID-significant2.sort.bam
$samtools index sampleID-significant2.sort.bam
$samtools idxstats sampleID-significant2.sort.bam >
sampleID_idxstats.txt
```

The generated file, sampleID_idxstats.txt, furnishes an exhaustive inventory of contigs and scaffolds, inclusive of their respective counts of mapped reads for each sample. Transform this text file into a .csv format, incorporating columns for Contig name, viral name, length, and the number of mapped reads. This resulting CSV file will be instrumental in subsequent processing steps.

3.2.6 Univirus

The objective of this stage is to identify shared viruses across all samples or pools. An R script will be employed to generate a list of viruses that are consistently present in all datasets. This process can

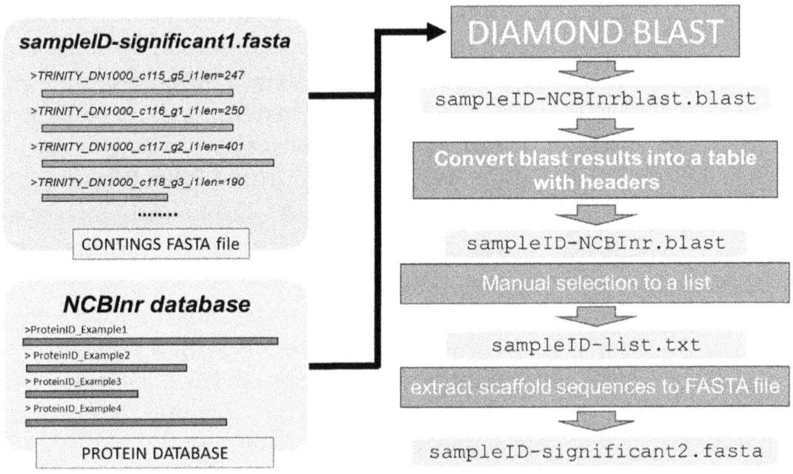

Fig. 6 Schematic representation of the use of DIAMOND software

be customized according to your preferences. You have the flexibility to modify this step, and you can create various tables or lists using R or RStudio. The corresponding R scripts are accessible on GitHub.

1. Rscript_mergeAllVOISequences.R (input: .csv → output: allVOI.fasta)

This script amalgamates all data into a single FASTA file with discernible names for traceability. The process involves the following steps:

1. Read data from .csv files in R, creating columns such as Contig name, viral name, length, and the number of mapped reads.
2. Consolidate virus names to assign the pool/sample name to each entry. All samples/pools will be unified into a single file.
3. Utilize Biostrings to read FASTA sequences. The script will leverage a FASTA file encompassing all contigs, subsequently performing a self-BLAST for comprehensive analysis.

2. blastallVOI.R (input: allVOI.fasta → output: allVOI_nucl.tab)

To execute this R script, the first step is to create a database file in the Unix command line for BLASTN analysis:

```
$makeblastdb -in allVOI.fasta -dbtype nucl
```

Subsequently, utilize the BLASTN tool to compare sequences in allVOI.fasta against the created database:

```
$blastn -query allVOI.fasta -db allVOI.fasta
-outfmt 6 -out allVOI_nucl.tab
```

With the generated allVOI_nucl.tab file, the identification of unique viruses will be conducted in R, employing an identity threshold. This process involves custom R scripting to compare the length of the contig with the alignment length. The alignment length is set to encompass at least 90% of the contig length, demonstrating a 90% identity. This meticulous comparison leads to the elimination of duplicate viruses, culminating in the creation of a list/table and a FASTA file exclusively comprising unique viruses.

3. Rscript_blastAnalysis.R (input: allVOI_nucl.tab → output: listvirusUni.txt)

With the execution of this R script, a comprehensive table summarizing the results, along with accompanying plots and graphics, can be generated. Subsequent to this step, proceed to extract univiruses from the FASTA file. Follow this by repeating specific BLAST steps against NCBI to assign names based on the first BLAST hit. Conclude this process with a new mapping, similar to the procedure outlined in step 3.2.5. Lastly, compile all pertinent data into a comprehensive table, forming an essential dataset to conclude the analysis of the mycovirome.

3.2.7 Molecular Analyses: ORF Predictions and Final BLASTP Against the NCBI Database

In this concluding phase, a comprehensive molecular analysis encompassing all identified viruses in our samples will be undertaken. For the ORF prediction of selected viral sequences, *Expasy* Translator and *Open Reading Frame Finder* from NCBI will be employed with default parameters. Subsequently, BLASTP will be utilized to validate the identity of the translated proteins by conducting another search against the NCBI database. This step is executable on a standard PC, and all tools will be accessed online.

Viral sequences meeting specific criteria, including a length exceeding 1000 nucleotides and possessing a complete coding sequence, notably encompassing proteins crucial for viral replication, will be earmarked for phylogenetic analyses and further mycovirome characterization. The entirety of this data should be promptly uploaded to a comprehensive database, such as GenBank-NCBI.

3.3 Design of Primers

A selection of mycoviruses with different genome types, dsRNA, ssRNA(+), ssRNA(−), and ssDNA (Table 1), belonging to different viral families, was performed to confirm mycoviral infection by PCR. Primer design was performed following the scheme in Fig. 7. Primers indicated in this chapter were used by Ruiz-Padilla and coworkers for the detection of the *B. cinerea* mycovirome [30].

1. Select mycoviruses for detection in fungal samples. Align mycoviral contigs belonging to the same genus or family using Clustal Omega. Use specific regions of the genome that do not show homology with the rest of the sequences belonging to mycoviruses of the same genus or family. In case of ssRNA(−) mycoviruses, sequences need to be reverse complemented to design primers.
2. Design primers in Primer3 web (Primer3 Input) (*see* **Notes 2 and 3**). In Primer3 software, paste specific regions of selected mycoviruses.
3. Select standard conditions for Primer Size, Primer Tm (melting temperature), Product Tm, and Primer GC% in *General Primer Picking conditions*, as shown in Fig. 8. Product size ranges were set between 80 and 100 for detection by qPCR.
4. Select primer pairs that show similar Tm values (*see* **Note 4**).
5. Paste primer sequences in Primer-BLAST (Primer designing tool (nih.gov)) to confirm that primers do not complement regions of the *B. cinerea* genome. Introduce primer sequences in *Primer Parameters* (Forward and Reverse). In *Primer Pair Specificity Checking Parameters*, select *Botrytis cinerea (taxid:40559)* (see Fig. 9).

3.4 Detection of Mycoviruses In Vivo by Quantitative PCR

1. Prepare a dilution of primers of 10 μM.
2. Synthesize cDNA from total RNA of each pool or field isolate using the High-Capacity cDNA Reverse Transcription

Table 1
List of primers used for the detection of different mycoviruses by RT-qPCR

Name	Type of genome	Forward primer (5′->3′)	Reverse primer (5′->3′)
Botrytis cinerea mycovirus 3 RNA1	dsRNA	GAGGACTTGGCTACGGTCAA	GCGTCTTGAGTCCCATCACT
Botrytis cinerea mycovirus 3 RNA2	dsRNA	CCATCTGGAAGACGGAGCTA	GACTGGCAGCATGTAAGTGG
Botrytis cinerea botybirnarvirus 2 RNA1	dsRNA	CGAAGTCGAGTAAGGGCAAG	ACACATCGCCCTCTTCAAAT
Botrytis cinerea botybirnarvirus 2 RNA2	dsRNA	CAACAACACTGGCAAACACC	TTGCCATCCATTCACGACTA
Botrytis porri botybirnavirus 1 RNA1	dsRNA	GCCTGTGGAAGCAGAAGAAG	CCTTCTACCAGGCTGTCTGC
Botrytis porri botybirnavirus 1 RNA2	dsRNA	CTGTCTACTCGGGCTCAAGG	GCAAATAGCAATGCAAACGA
Botrytis cinerea mycovirus 5 RNA1	dsRNA	CTCAACGCCAAGTACGACAA	AGCTTCTCATCGAGCCTCAC
Botrytis cinerea mycovirus 5 RNA2	dsRNA	CGGCTTCAGGACTTCGATAA	ATCCAATGTCGGCATAGAGC
Botryotinia fuckeliana partitivirus 1 RNA1	dsRNA	AGGTCGTTTCAATCCAACCA	TCGTCACGTTCTGACATTCC
Botryotinia fuckeliana partitivirus 1 RNA2	dsRNA	ACGGGTAAGACAGGATGTGC	CACCTTTTGTGTGTCCGTCA
Botrytis cinerea mycovirus 4 RNA1	dsRNA	GCAGGCACGTTACACAAATG	GCACCTCGTGTGATTGCTTA
Botrytis cinerea mycovirus 4 RNA2	dsRNA	ATGCGCAAGGTAGTTTCACC	CTTTGGTTTGGGGCATAAGA
Botrytis cinerea mycovirus 4 RNA3	dsRNA	GATCCCGTAACAACGCAAGT	TGCCCAAGATGTTCCACATA
Botrytis cinerea mycovirus 4 RNA4	dsRNA	GCATGCACCGAGAAGACATA	GGCATCGAATCTGGTCACTT
Botrytis cinerea victorivirus 2_BCS9	dsRNA	TGGCTCACTGTTGGTTGAAG	TGACCGTATCTCGTTGGTCA
Botrytis cinerea victorivirus 2_BCS14	dsRNA	ACACCCTTAGTGACGCTGCT	TTGTTTTGTGGGGTTCATCC
Botrytis cinerea victorivirus 3	dsRNA	GAGCGATATTCGCCTGTGAT	CCACCTCTTCTGGACGTCAT
Botrytis cinerea alpha-like virus 1	ssRNA(+)	CTGTGTCCGGCACTCAATTA	TCCATAACGCCAACAATTCA
Botrytis cinerea flexivirus 1	ssRNA(+)	ACTCCTCGCCTGCTGATAGA	ATCGGTATACAGGCCACTCG
Botrytis cinerea deltaflexivirus 1	ssRNA(+)	CAAGGCAACGTTTTTCTCGT	TGACTGTTGCCATACGTGGT
Botrytis cinerea umbra-like virus 1	ssRNA(+)	CAGCGAGTCATTTGTCAGGA	ACGGTCAAAGCTACGTCCTC
Botrytis cinerea endornavirus 2	ssRNA(+)	TGTCGAGTACCCCATGTTCA	GGGTCCTATCCAAACCAACA
Botrytis cinerea endornavirus 3	ssRNA(+)	GGCAGGTGTGACAAAAGGTT	GAAAGGCAGTGCAAATGGTT

Botrytis cinerea fusarivirus 3	ssRNA(+)	CTCGTGACAGACGCACAAGT	GGGGACCGTGCCTCTATATT
Botrytis cinerea fusarivirus 5	ssRNA(+)	CACGTGCAGTCGATCATCTT	CCGATGTCTTGCTCATTGAA
Botrytis cinerea fusarivirus 7	ssRNA(+)	GCAGAATCACACCCATCAGA	TTTGCAGAAGGAGCAAAAGG
Botrytis cinerea hypovirus 2	ssRNA(+)	GCATGATGAAGAGCCAGTGA	GGGCTATCACCAAATTCCAA
Botrytis cinerea hypovirus 3	ssRNA(+)	CGACGCCCTTTAAAATACCA	TGCGTCATACAAGCAGGAAG
Botrytis cinerea hypovirus 4	ssRNA(+)	GGACGAGGACACACTTTGGT	GAACGCAGGTGTTGCAGTTA
Botrytis cinerea binarnavirus 1	ssRNA(+)	ATGGAAGATCCATGCTGCTC	GCCAGACCACTAGCTCAGGA
Botrytis cinerea binarnavirus 2 RdRp	ssRNA(+)	ACGGAAGTGGTTCAAGATGC	CGATAGGTATGTCGCCGTCT
Botrytis cinerea binarnavirus 2 HP	ssRNA(+)	AGAGTCCCCACTCTCCGTCT	CCGTATTCCATGTCCTCGTC
Botrytis cinerea narnavirus 4	ssRNA(+)	GAGCCCTTCTCCAGAAATCC	AGCTTCACGAATGGCATT
Botrytis cinerea binarnavirus 5	ssRNA(+)	CCTTCAACGGTTCCTCTCAT	GAAATGCCCTGAAGGACAAA
Botrytis cinerea ourmia-like virus 9	ssRNA(+)	CTGGCTTCTTGTTGGTAGCC	GAGTGCCACACCTTTCCTTC
Botrytis cinerea ourmia-like virus 11	ssRNA(+)	CCTTTCTCTGCCTGCTCAAT	CCTACACGCCTTTCACCTTC
Botrytis cinerea ourmia-like virus 16	ssRNA(+)	GCTTACGGATTTGGAACGAG	GGCCAGAGTCCTGCAAGTTA
Sclerotinia sclerotiorum mitovirus 4	ssRNA(+)	GGCTCATGCAACGACAAGAT	GGTCAACCAATGGGAGCTTA
Botrytis cinerea mitovirus 6	ssRNA(+)	CAATGCCTCAGTGATCGAAA	TACCTTTCCTGCTGGCTCAT
Botrytis cinerea mitovirus 5	ssRNA(+)	TGGACAACCTATGGGAGCTT	TGCGGCTCATGCAACTACTA
Botrytis cinerea mitovirus 9	ssRNA(+)	TGGGATTTAACCGCAAATCA	ACTATTCGGACCAGCAGCTT
Botrytis cinerea bocivirus-RNA1	ssRNA(−)	CAGCAGATGCAGCTAAGTGG	GGATGGGAGAAATGAAGCAA
Botrytis cinerea bocivirus-RNA2	ssRNA(−)	TCCCTATGGAAAAGCTGTGG	CATGAAGACTTCCCTGCAAGA
Botrytis cinerea bocivirus-RNA3	ssRNA(−)	GCATGTGCACCATTGCTATC	GACCTCTTTCCAGCGTCTGA
Botrytis cinerea orthobunya-like virus 1	ssRNA(−)	CAGCACTTTCGTGTCCAAAA	TTTGCAAGCCTTAGGAGACA
Botrytis cinerea negative-stranded RNA virus 3	ssRNA(−)	CGGCATTCCAATCGCTATAC	CGCAGATCGAGAAAGAATCC
Botrytis cinerea negative-stranded RNA virus 6	ssRNA(−)	GATGTTCCACAAGCTCAGCA	CACGTCCTTGATCACCTCCT
Botrytis cinerea ssDNA virus 1	ssDNA	GTCGACGAATTTTTGGAGGA	GAAGGCTTTGAATTGGGTGA

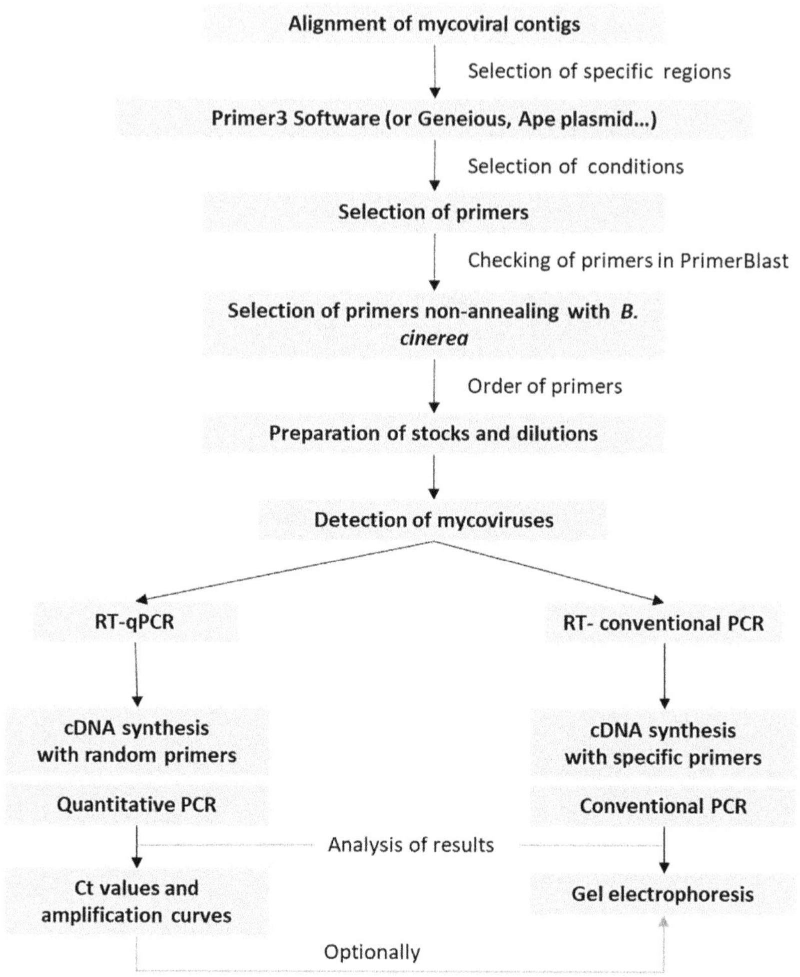

Fig. 7 Workflow scheme followed for the detection of mycoviruses in fungal samples, from primer design to PCR amplification

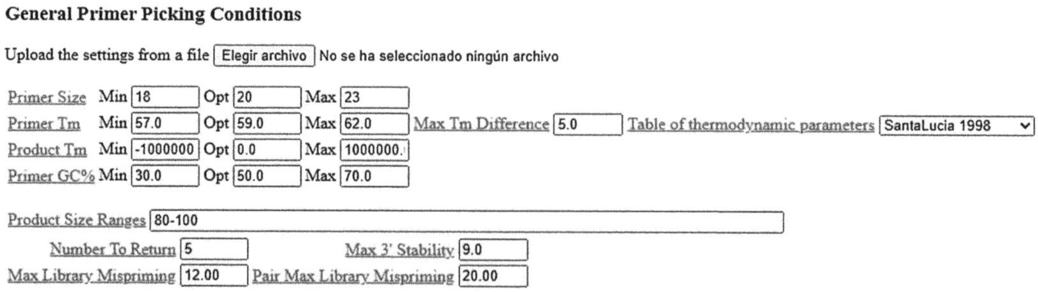

Fig. 8 Primer design parameters applied in the Primer3 website

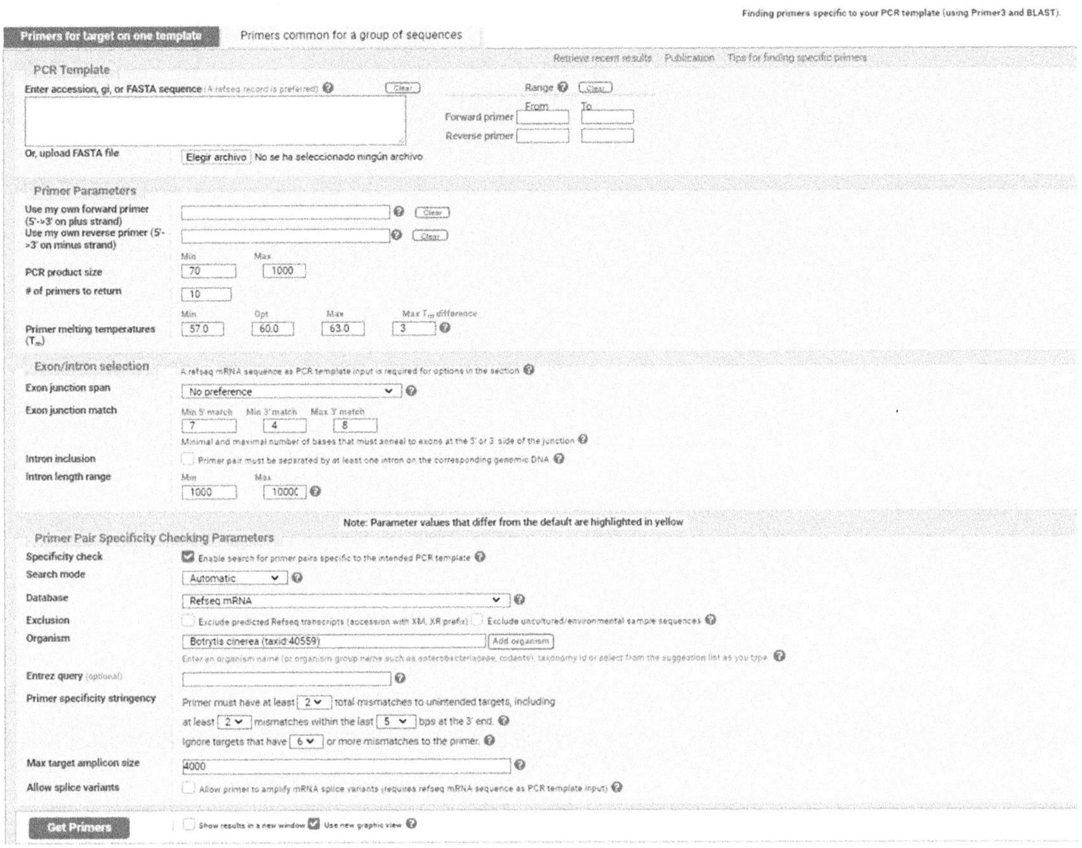

Fig. 9 Parameters used for primer testing against Botrytis cinerea (taxid:40559) in Primer-BLAST

Kit. Mix 1 μg of total RNA with 1× RT Buffer, 4 μM of dNTP Mix, 1× Random Primers, and 25 U of MultiScribe™ Reverse Transcriptase.

3. Incubate samples at 25°C for 10 min, 37°C for 120 min, and 85°C for 5 min, then cool at 4°C.

4. Prepare qPCR reactions using reagents of FastStart Universal SYBR Green Master Rox (Roche). Prepare a Master Mix with 1× SYBR Green Master (ROX) with 20 μM of primers and nuclease-free water in a total volume of 10 μL per single reaction (*see* **Note 5**). Load Master Mix in 96-well or a 384-well plates using an automatic pipette (*see* **Notes 6 and 7**).

5. Load 1 μL of 1/3 diluted cDNA into each well, using two technical replicates per sample and primer pair.

6. Seal the plate and centrifuge for 30 s at 1107 × *g*. Introduce the plate into the Light Cycler 480 using the following program (*see* **Note 8**).

7. Analyze results considering positive Ct values lower than 30 (*see* **Note 9**). In Table 2, examples of mycoviruses detected by RT-qPCR are shown.

	# Cycles	Temperature (°C)	Time (mm:ss)	Analysis mode
Preincubation	1	95	05:00	None
Amplification	45	95	00:10	None
		53	00:20	None
		72	00:30	Single
Melting curves	1	95	00:05	None
		65	01:00	None
		97		Continuous
Cooling	1	40	00:10	None

3.5 Detection of Mycoviruses by Conventional PCR

Mycoviruses can also be detected by reverse transcription using specific primers and conventional PCR. In this case, primers are designed following the same procedure as explained in Sect. 3.3. with the modification of the target size to 300–500 bp.

1. Prepare a dilution of reverse primers of 2 μM.
2. For reverse transcription, prepare annealed RNA by mixing 200 ng of total RNA from a pool or field isolate with 1 μL of 10 mM dNTP mix, 1 μL of 2 μM reverse primer, and complete with nuclease-free water to a final reaction volume of 20 μL.
3. Use the following program in the thermal cycler or use a thermal block.

# Cycle	Temperature (°C)	Time (mm:ss)
1	65	05:00
	4 (ice)	At least 01:00

4. Prepare Master Mix with 4 μL of 5× SSIV Buffer, 1 μL of 100 mM DTT, 1 μL of RNAse inhibitor, and 1 μL of SuperScript IV®.
5. Mix the RT reaction with annealed RNA from step 3 and incubate samples in a thermal cycler using the following program (*see* **Note 10**).

# Cycle	Temperature (°C)	Time (mm:ss)
1	50	60:00 (*see* **Note 11**)
	80	10:00
	4	Infinite

6. Optionally, RNA can be removed using 1 μL of *E. coli* RNase H and incubating at 37°C for 20 min.

Table 2
Quantitative PCR results of the detection of selected mycoviruses

<table>
<tr><th>Mycovirus</th><th>Genome type</th><th>Amplification curves</th><th>Ct values obtained</th></tr>
<tr><td>Botrytis cinerea mycovirus 3 RdRp</td><td>dsRNA</td><td>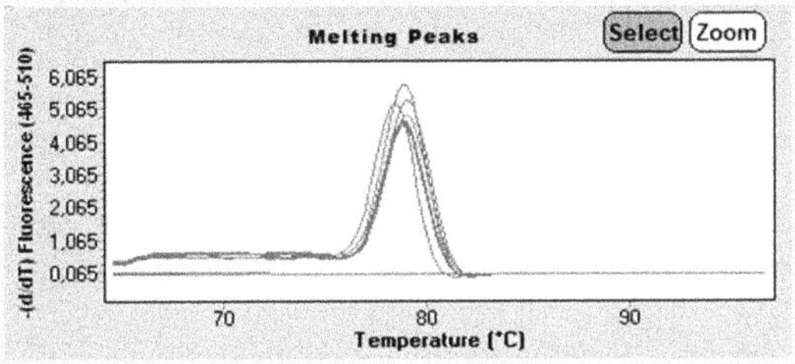
</td><td>22.59
21.78
21.80
22.29
27.63
28.61
22.82
22.39</td></tr>
<tr><td>Botrytis cinerea mycovirus 3 HP</td><td>dsRNA</td><td>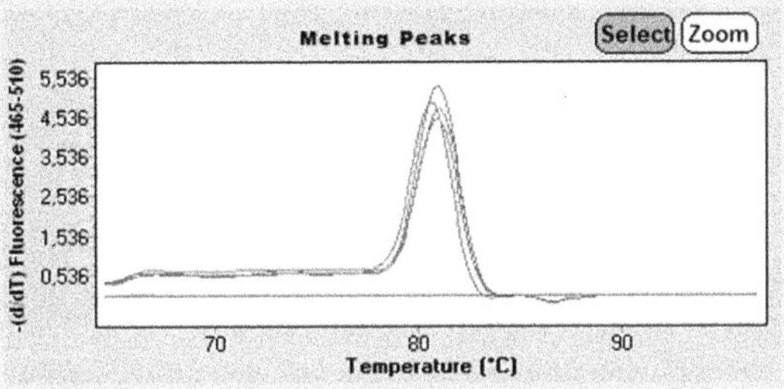
</td><td>22.49
22.60
20.87
20.70
22.40
22.65
22.51
22.47</td></tr>
<tr><td>Botrytis cinerea hypovirus 2</td><td>ssRNA(+)</td><td>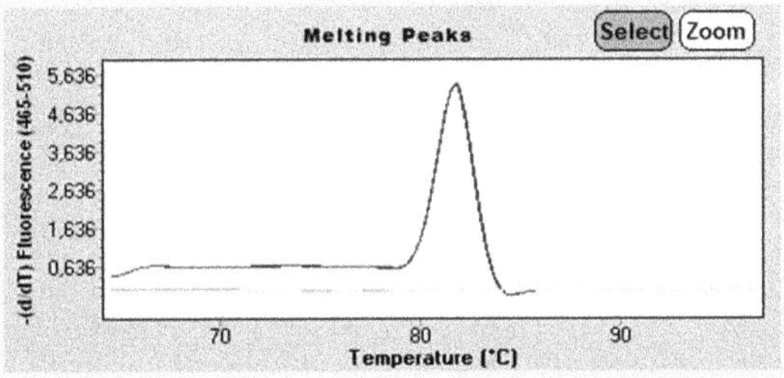
</td><td>23.45
22.53
19.62
19.46</td></tr>
<tr><td>Botrytis cinerea bocivirus 1 RNA1</td><td>ssRNA(−)</td><td>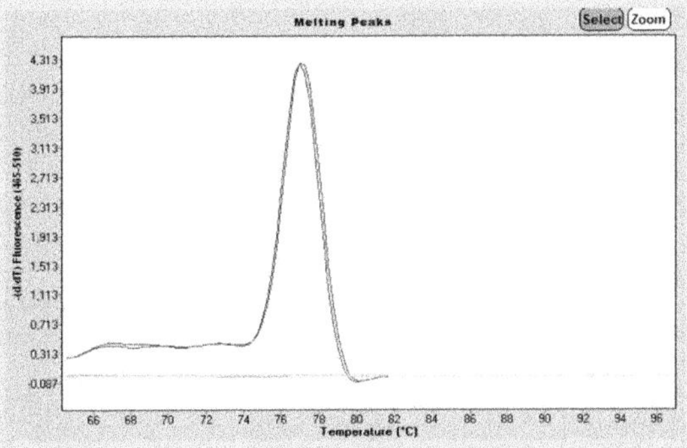
</td><td>21.37
21.53</td></tr>
</table>

(continued)

Table 2
(continued)

Mycovirus	Genome type	Amplification curves	Ct values obtained
Botrytis cinerea bocivirus 1 RNA2	ssRNA(−)	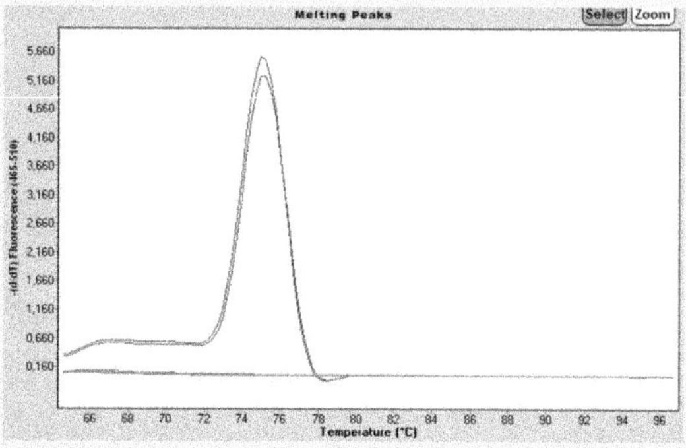	27.24 27.59
Botrytis cinerea bocivirus 1 RNA3	ssRNA(−)	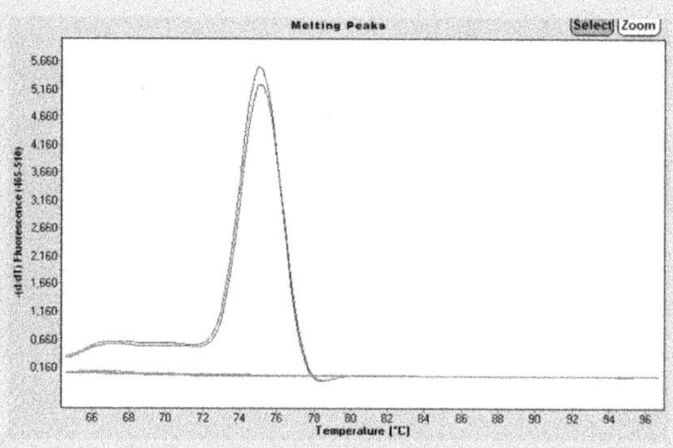	20.73 20.83
Botrytis cinerea ssDNA virus 1	ssDNA	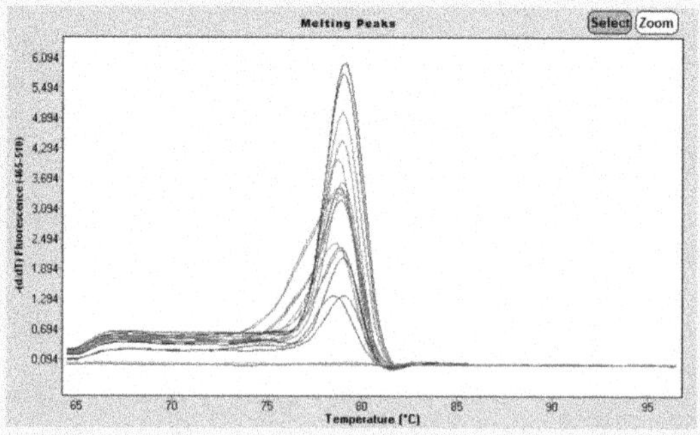	25.78 24.35 29.95 28.63

Columns 3 and 4 show amplification curves of positive samples and negative controls (plain lines) and Ct values of some samples (two technical replicates per sample)

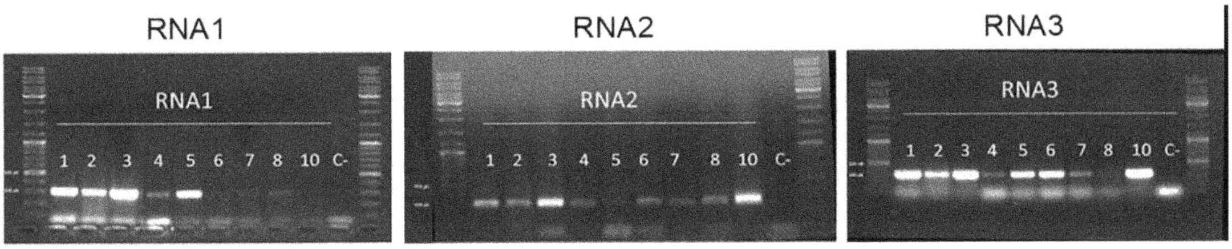

Fig. 10 PCR detection of RNA segments 1, 2, and 3 of *Botrytis cinerea* bocivirus 1, with expected amplicon sizes of ~300 bp

7. Perform PCR using Taq CloneAmp HiFi PCR (TaKaRa) following the manufacturer's instructions. Use 1 μl of 1/3 diluted cDNA in a total reaction volume of 25 μL. For maximum efficiency, use specific melting temperatures.

An example of positive detection of three segments of *Botrytis cinerea* bocivirus 1 is shown in Fig. 10.

3.6 Validation of Independence of Segments in Multisegmented Mycoviruses

Some mycoviruses can be multisegmented, and molecular analysis can confirm the independence of the segments.

3.6.1 Determination of Mycoviral Genomic 5′ and 3′ Ends

1. Design two primers for each genomic end. For the 5′ end, outer and inner primers are located at 150 and 200 nucleotides from the terminal nucleotide, and both are in antisense orientation. For the 3′ end, the same design is applied, with both primers in the sense orientation (*see* **Note 12** and Fig. 11).
2. Use the SMARTer RACE Kit or RLM-RACE Kit to determine the 5′ and 3′ ends (*see* **Notes 13 and 14**), following the manufacturer's instructions.

3.6.2 Combined PCRs

1. Design primers PS1 (forward) and PS2 (reverse) to amplify approximately from the half part of genomic segment 1 to the half part of genomic segment 2 (1 kb), and primers PS3 (forward) and PS4 (reverse) to amplify from the half part of genomic segment 2 to the half part of genomic segment 1. Design primers PD1 and PD2 to amplify 1 kb of genomic segment 1 and PD3 and PD4 to amplify 1 kb of genomic segment 2 (see Fig. 12). Design primers following the instructions in Sect. 3.3, selecting specific regions as in Fig. 12.
2. Use cDNA synthesized with random primers, as described in Sect. 3.4.1, and perform PCR with CloneAmp Taq Polymerase using the next combinations of primers: PS1 and PS2, PD1 and PD2, and PD3 and PD4 (*see* **Note 15**).

Fig. 11 Primer positions for amplification of the 5′ and 3′ genomic ends

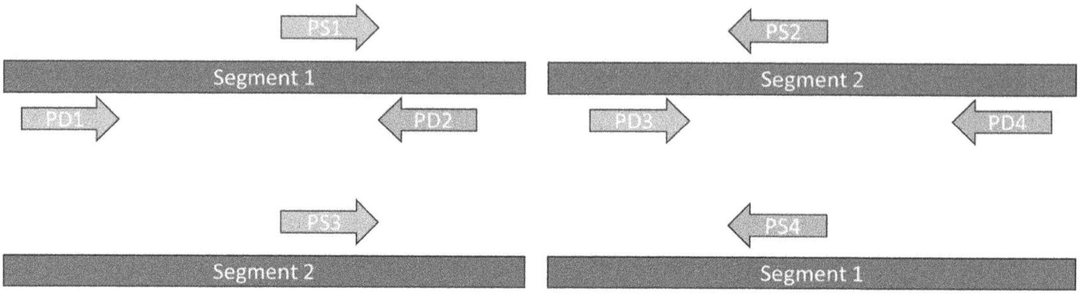

Fig. 12 Primer positions in genomic segments 1 and 2 to confirm the independence of the segments

3. To validate the independence of the segments, amplifications with primers PS1 and PS2, and PS3 and PS4, must be negative. Control amplifications with PD1 and PD2, and PD3 and PD4, must be positive. An example is shown in Fig. 13.

3.7 In Silico Search of Additional Segments in Multisegmented Genomes

Usually, the most frequently detected genomic segment in silico is the one containing the open reading frame coding for the RNA-dependent RNA polymerase. Additional segments of mycoviruses with multisegmented genomes can be detected by analysis of contigs using nucleotide or protein sequences as reference in a local BLASTN or BLASTX search.

3.7.1 Analysis Using Nucleotide Sequences

1. Use the *cd* command to move to the location of files.
   cd./working_directory/
2. Move the sample_trinity.fasta file from directory 1 to the working directory using the *mv* command.
   mv./directory1/sample_trinity.fasta./working_directory
3. Prepare a BLAST database with a FASTA file including nucleotide sequences to be used as reference. Open a file in Atom or Notepad and follow the structure below. Save the file as database.fasta.
   >Virus1
   Nucleotide sequence
   Virus2
   Nucleotide sequence

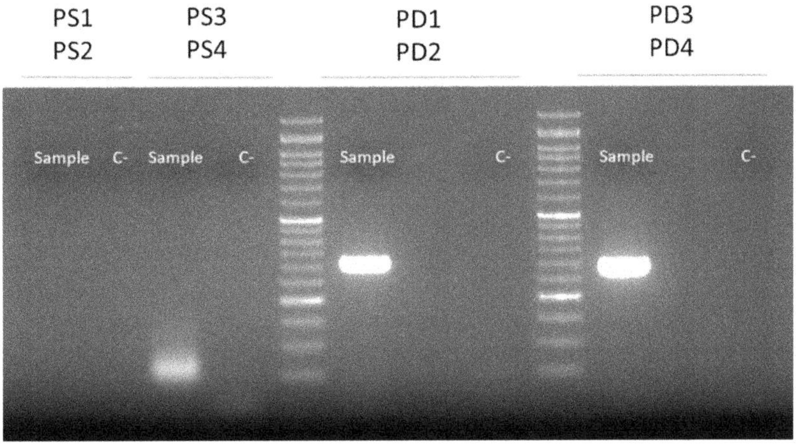

Fig. 13 PCR results obtained with primer combinations PS1–PS2, PS3–PS4, PD1–PD2, and PD3–PD4 to confirm the independence of the segments

4. Open Ubuntu Desktop in Windows or the terminal in Ubuntu.
5. Use the *cd* command to move to the location of files if necessary.
6. Insert command: makeblastdb -in./working_directory/database.fasta -dbtype nucl -title database
7. Insert command: blastn -query./working_directory/sample_Trinity_file.fasta -db./working_directory/database.fasta -out blastn.tab -evalue 0.01 -outfmt ' 6 qseqid sseqid pident length qlen mismatch gapopen qstart qend sstart send evalue bitscore stitle' -num_descriptions 2 (*see* **Notes 16–18**).
8. Open the resulting blastn.tab file and copy the table into an Excel file.
9. Order results by length and identity to identify putative additional segments.
10. Select putative viral contigs from the FASTA file.

3.7.2 Analysis Using Protein Sequences

For protein sequences, similar steps are carried out, substituting the commands in Sect. 3.7.1 with *makeblastdb -in./working_directory/database.fasta -dbtype prot -title database* and *blastx -query./working_directory/sample_Trinity_file.fasta -db./db/database.fasta -out blastn.tab -evalue 0.01 -outfmt ' 6 qseqid sseqid pident length qlen mismatch gapopen qstart qend sstart send evalue bitscore stitle' -num_descriptions 2*.

4 Notes

1. For total RNA extraction, mycelia with liquid nitrogen can also be used with NZYol (NZYTech) or TRIzol (Invitrogen). For optimal RNA quality and quantity, and for a high number of samples, the Spectrum Plant Total RNA Kit is recommended.
2. Alternatively, primers can be designed with software such as Geneious or Ape plasmid, selecting similar product size ranges.
3. In the case of multisegmented viruses, primers were designed for each segment.
4. It is recommended that all primers selected for validation have similar melting temperatures in order to include all reactions in the same 96-well or 384-well plates.
5. Calculations of Master Mixes should be done with at least +2 reactions per specific primer pair.
6. In case of not using an automatic pipette, make sure loaded volumes are equal for each well.
7. Use filter tips to avoid contamination.
8. Amplification cycles can be reduced to 35, but low viral titer of certain viruses makes it necessary to add ten more cycles.
9. In case of high Ct values, an electrophoresis gel (2%) can be run to visualize bands and confirm positive results.
10. Alternatively, thermal blocks can be used, but temperature changes should be as fast as possible. Also, chilling at 4°C must be performed to avoid degradation of complementary DNA.
11. Incubation time can be minimized to 10 min, depending on the expected amplicon lengths for PCR.
12. It is important to carefully read and follow the manufacturer's instructions to design primers for determination of 5′ and 3′ genomic ends.
13. Better amplification results of 5′ terminal ends are obtained with the SMARTer RACE Kit.
14. The initial amount of total RNA must be maximized to determine mycoviral ends.
15. Alternatively, cDNA can be synthesized with specific primers using SuperScript IV or other reverse transcriptases.
16. The E-value parameter can be modified by increasing it or decrease it depending on the similarity expected to be found.
17. The number of descriptions parameter can be increased to obtain more results per analyzed contig.

18. In case of searching multiple files, a loop can be included, resulting in the following command, where 1, 2, and 3 are the changing names of the Trinity files: *for index in 1 2 3 4; do blastn -query.${index}_trinity.fasta -db database.fasta -out ${index}.tab -evalue 0.01 -outfmt ' 6 qseqid sseqid pident length qlen mismatch gapopen qstart qend sstart send evalue bitscore stitle'; done.*

Acknowledgments

This work was supported by the European Union's Horizon 2020 Research and Innovation Program under Grant Agreement No. 773567. M. A. Ayllón and J.L.R-R research is currently funded by the Project I+D+i PID2020-120106RB-I00 supported by MCIN/ AEI/10.13039/501100011033/

References

1. Fillinger S, Elad Y (2016) Botrytis – the fungus, the pathogen and its management in agricultural systems. Springer International Publishing, Cham
2. Dean R, Van Kan JAL, Pretorius ZA et al (2012) The top 10 fungal pathogens in molecular plant pathology. Mol Plant Pathol 13:804. https://doi.org/10.1111/j.1364-3703.2012.00822.x
3. Fernández-Ortuño D, Pérez-García A, Chamorro M et al (2017) Resistance to the SDHI fungicides boscalid, fluopyram, fluxapyroxad, and penthiopyrad in *Botrytis cinerea* from commercial strawberry fields in Spain. Plant Dis 101:1306–1313. https://doi.org/10.1094/PDIS-01-17-0067-RE
4. Fernández-Ortuño D, Grabke A, Li X, Schnabel G (2015) Independent emergence of resistance to seven chemical classes of fungicides in *Botrytis cinerea*. Phytopathology 105:424–432. https://doi.org/10.1094/PHYTO-06-14-0161-R
5. Fernández-Ortuño D, Torés JA, Chamorro M et al (2016) Characterization of resistance to six chemical classes of site-specific fungicides registered for gray mold control on strawberry in Spain. Plant Dis 100:2234–2239. https://doi.org/10.1094/PDIS-03-16-0280-RE
6. Márquez LM, Redman RS, Rodriguez RJ, Roossinck MJ (2007) A virus in a fungus in a plant: three-way symbiosis required for thermal tolerance. Science 315:513–515. https://doi.org/10.1126/science.1136237
7. Zhang H, Xie J, Fu Y et al (2020) A 2-kb mycovirus converts a pathogenic fungus into a beneficial endophyte for Brassica protection and yield enhancement. Mol Plant 13:1420–1433. https://doi.org/10.1016/j.molp.2020.08.016
8. Ahn I-P, Lee Y-H (2001) A viral double-stranded RNA up regulates the fungal virulence of *Nectria radicicola*. Mol Plant Microbe Interact 14:496–507. https://doi.org/10.1094/MPMI.2001.14.4.496
9. Shah UA, Kotta-Loizou I, Fitt BDL, Coutts RHA (2020) Mycovirus-induced hypervirulence of *Leptosphaeria biglobosa* enhances systemic acquired resistance to *Leptosphaeria maculans* in *Brassica napus*. Mol Plant Microbe Interact 33:98–107. https://doi.org/10.1094/MPMI-09-19-0254-R
10. Olivé M, Campo S (2021) The dsRNA mycovirus ChNRV1 causes mild hypervirulence in the fungal phytopathogen *Colletotrichum higginsianum*. Arch Microbiol 203:241–249. https://doi.org/10.1007/s00203-020-02030-7
11. Córdoba L, Ruiz-Padilla A, Rodríguez-Romero J, Ayllón MA (2022) Construction and characterization of a *Botrytis* virus F infectious clone. J Fungi 8:459. https://doi.org/10.3390/jof8050459
12. Córdoba L, Ruiz-Padilla A, Pardo-Medina J, Rodríguez-Romero JL, Ayllón MA (2024) Construction of a Mycoviral Infectious Clone for Reverse Genetics in Botrytis cinerea. Methods Mol Biol 2751:47–68. https://doi.org/10.1007/978-1-0716-3617-6_4
13. Hollings M (1962) Viruses associated with a die-back disease of cultivated

mushroom. Nature 196:962–965. https://doi.org/10.1038/196962a0
14. Rodríguez-García C, Medina V, Alonso A, Ayllón MA (2014) Mycoviruses of *Botrytis cinerea* isolates from different hosts. Ann Appl Biol 164:46–61. https://doi.org/10.1111/aab.12073
15. Dawe AL, Nuss DL (2013) Hypovirus molecular biology: from Koch's postulates to host self-recognition genes that restrict virus transmission. Adv Virus Res 86:109–147. https://doi.org/10.1016/B978-0-12-394315-6.00005-2
16. Rigling D, Prospero S (2018) *Cryphonectria parasitica*, the causal agent of chestnut blight: invasion history, population biology and disease control. Mol Plant Pathol 19:7–20. https://doi.org/10.1111/mpp.12542
17. García-Pedrajas MD, Cañizares MC, Sarmiento-Villamil JL et al (2019) Mycoviruses in biological control: from basic research to field implementation. Phytopathology 109:1828–1839. https://doi.org/10.1094/PHYTO-05-19-0166-RVW
18. Khan HA, Mukhtar M, Bhatti MF (2023) Mycovirus-induced hypovirulence in notorious fungi *Sclerotinia*: a comprehensive review. Braz J Microbiol 54:1459–1478. https://doi.org/10.1007/s42770-023-01073-4
19. Wagemans J, Holtappels D, Vainio E et al (2022) Going viral: virus-based biological control agents for plant protection. Annu Rev Phytopathol 60:21–42. https://doi.org/10.1146/annurev-phyto-021621-114208
20. Donaire L, Rozas J, Ayllón MA (2016) Molecular characterization of *Botrytis ourmia*-like virus, a mycovirus close to the plant pathogenic genus *Ourmiavirus*. Virology 489:158–164. https://doi.org/10.1016/j.virol.2015.11.027
21. Galán-Cubero R, Córdoba L, Rodríguez-Romero J et al (2022) Molecular data of a novel penoulivirus associated with the plant-pathogenic fungus *Erysiphe necator*. Phytopathology. https://doi.org/10.1094/PHYTO-12-21-0536-A
22. Ruiz-Padilla A, Turina MAM (2023) Molecular characterization of a tetra segmented ssDNA virus infecting *Botrytis cinerea* worldwide. Virol J 20:306
23. Ruiz-Padilla A, Rodríguez-Romero JL, Pacifico D, Chiapello MAM (2024) Determination of the mycovirome of a necrotrophic fungus. Methods Mol Biol 2732:83–101
24. Donaire L, Pagán I, Ayllón MA (2016) Characterization of *Botrytis cinerea* negative-stranded RNA virus 1, a new mycovirus related to plant viruses, and a reconstruction of host pattern evolution in negative-sense ssRNA viruses. Virology 499:212–218. https://doi.org/10.1016/j.virol.2016.09.017
25. Donaire L, Ayllón MA (2017) Deep sequencing of mycovirus-derived small RNAs from *Botrytis* species. Mol Plant Pathol 18:1127–1137. https://doi.org/10.1111/mpp.12466
26. Illana A, Marconi M, Rodríguez-Romero J et al (2017) Molecular characterization of a novel ssRNA ourmia-like virus from the rice blast fungus *Magnaporthe oryzae*. Arch Virol 162:891–895. https://doi.org/10.1007/s00705-016-3144-9
27. Nerva L, Forgia M, Ciuffo M et al (2019) The mycovirome of a fungal collection from the sea cucumber *Holothuria polii*. Virus Res 273: 197737. https://doi.org/10.1016/j.virusres.2019.197737
28. Sutela S, Forgia M, Vainio EJ et al (2020) The virome from a collection of endomycorrhizal fungi reveals new viral taxa with unprecedented genome organization. Virus Evol. https://doi.org/10.1093/ve/veaa076
29. Crucitti D, Chiapello M, Oliva D et al (2021) Identification and molecular characterization of novel mycoviruses in *Saccharomyces* and non-*Saccharomyces* yeasts of oenological interest. Viruses 14:52. https://doi.org/10.3390/v14010052
30. Ruiz-Padilla A, Rodríguez-Romero J, Gómez-Cid I, Pacifico D, Ayllón MA (2021) Novel mycoviruses discovered in the mycovirome of a necrotrophic fungus. MBio 12:e03705–e03720
31. Botella L, Jung T (2021) Multiple viral infections detected in *Phytophthora condilina* by total and small RNA sequencing. Viruses 13:620. https://doi.org/10.3390/v13040620
32. Khalifa ME, MacDiarmid RM (2021) A mechanically transmitted DNA mycovirus is targeted by the defence machinery of its host, *Botrytis cinerea*. Viruses 13:1315. https://doi.org/10.3390/v13071315
33. Hao F, Wu M, Li G (2021) Characterization of a novel genomovirus in the phytopathogenic fungus *Botrytis cinerea*. Virology. https://doi.org/10.1016/j.virol.2020.11.007
34. Yu X, Li B, Fu Y et al (2010) A geminivirus-related DNA mycovirus that confers hypovirulence to a plant pathogenic fungus. Proc Natl Acad Sci U S A 107:8387–8392. https://doi.org/10.1073/pnas.0913535107
35. Li P, Wang S, Zhang L et al (2020) A tripartite ssDNA mycovirus from a plant pathogenic fungus is infectious as cloned DNA and purified virions. Sci Adv 6:eaay9634. https://doi.org/10.1126/sciadv.aay9634
36. Ayllón MA, Vainio EJ (2023) Mycoviruses as a part of the global virome: diversity, evolutionary links and lifestyle. Adv Virus Res 115:1–86. https://doi.org/10.1016/bs.aivir.2023.02.002

37. Grabherr MG, Haas BJ, Yassour M et al (2011) Full-length transcriptome assembly from RNA-Seq data without a reference genome. Nat Biotechnol 29:644–652. https://doi.org/10.1038/nbt.1883
38. Bushnell B, Rood J, Singer E (2017) BBMerge – accurate paired shotgun read merging via overlap. PLoS One 12:e0185056. https://doi.org/10.1371/journal.pone.0185056
39. Buchfink B, Xie C, Huson DH (2015) Fast and sensitive protein alignment using DIAMOND. Nat Methods 12:59–60. https://doi.org/10.1038/nmeth.3176
40. Danecek P, Bonfield JK, Liddle J et al (2021) Twelve years of SAMtools and BCFtools. Gigascience 10. https://doi.org/10.1093/gigascience/giab008
41. Li H, Durbin R (2009) Fast and accurate short read alignment with Burrows-Wheeler transform. Bioinformatics 25:1754–1760. https://doi.org/10.1093/bioinformatics/btp324
42. Huang X, Madan A (1999) CAP3: a DNA sequence assembly program. Genome Res 9: 868–877. https://doi.org/10.1101/gr.9.9.868
43. Milne I, Stephen G, Bayer M et al (2013) Using Tablet for visual exploration of second-generation sequencing data. Brief Bioinform 14:193–202. https://doi.org/10.1093/bib/bbs012
44. Altschul SF, Gish W, Miller W et al (1990) Basic local alignment search tool. J Mol Biol 215:403–410. https://doi.org/10.1016/S0022-2836(05)80360-2
45. Chiapello M, Rodríguez-Romero J, Ayllón MA, Turina M (2020) Analysis of the virome associated to grapevine downy mildew lesions reveals new mycovirus lineages. Virus Evol. https://doi.org/10.1093/ve/veaa058

Chapter 12

Botrytis cinerea Protoplasts Production and Inoculation

Javier Pardo-Medina, Julián Méndez-García, Laura Córdoba and María A. Ayllón

Abstract

The absence of a cell wall turns fungal protoplasts very receptive to the uptake of exogenous genetic material and, consequently, they can be used for genetic transformation or transfection experiments. Additionally, isolated monosporic fungal strains, naturally infected or previously transfected with a mycovirus, can be used in assays of fungal protoplasts fusion. This technology will overcome the barriers of vegetative incompatibility with other *Botrytis cinerea* strains, or even with other fungi, and successfully allow the transference of mycoviruses from an infected to a non-infected fungal strain. We describe in this chapter the protocols to isolate *B. cinerea* protoplasts; to perform fungal protoplasts transfection or transformation; and to do protoplasts fusion assays to transfer mycoviruses from infected to non-infected *B. cinerea* strains.

Keywords Mycovirus, Protoplast, Transformation, Transfection, Mycoviral infectious clone, Reverse genetics, Fungus, *Botrytis cinerea*

1 Introduction

Significant advances in fungal and mycoviral research fields have been achieved through experiments performed in fungal protoplasts. A large number of assays have been carried out using plasmids as input for fungal protoplasts transformation. Additionally, to understand mycoviral life cycle inside a fungus, its transmission to other fungal strains, or its effect on fungal pathogenicity, analyses have been developed with fungal protoplasts using mycoviral infectious full-length clones. These synthetic mycoviruses have been constructed based on DNA mycoviruses using recombinant plasmids for fungal protoplast transformation. This strategy has been followed in the case of two single-stranded DNA mycoviruses: Fusarium graminearum gemytripvirus 1 [1] and soybean-leaf-associated gemycircularvirus 1 [2]. Furthermore, the construction of RNA mycoviruses has been mostly based on the cloning of the genomic cDNA copy in an appropriate plasmid, followed by *in vitro* synthesis of RNA transcripts,

Luca Nerva and Walter Chitarra (eds.), *Botrytis cinerea*, Methods in Molecular Biology, vol. 3076,
https://doi.org/10.1007/978-1-0716-5551-1_12,

and transfection of fungal protoplasts. This last approach has been executed in the case of Cryphonectria hypovirus [3], Saccharomyces 23S RNA narnavirus [4], Saccharomyces 20S RNA narnavirus [4], Diaporthe RNA virus 1 [5], Sclerotinia sclerotiorum hypovirus 2 Lactuca [6], yado-kari virus 1 (YkV1) and yado-nushi virus 1 (YnV1) [7], Sclerotinia sclerotiorum ourmia-like virus 4 (SsOLV4) [8], and Botrytis virus F (BVF) [9, 10].

Over a hundred mycoviruses have been identified in the plant pathogenic fungus *Botrytis cinerea* [11–21]. BVF is the only mycovirus infecting *B. cinerea* used as a template to construct a successful infectious clone to study its effect on the phenotype and virulence of different *B. cinerea* strains and different plant hosts [9]. BFV has a monosegmented positive single-stranded RNA genome with a poly-A tail and two monocistronic open reading frames [9, 22–24]. Using this system, under *in vitro* conditions, it was shown that BVF has no effect on the growth rate of two different *B. cinerea* strains. Additionally, it was demonstrated that BVF had no effect on the virulence of a laboratory strain of *B. cinerea* but induced hypovirulence in a field strain infected with two other mycoviruses. Then, the authors of this work proposed the putative use of BVF in biocontrol strategies. The success of this study was related to an excellent system of production and transfection of fungal protoplasts. In this chapter, we describe the detailed protocols to obtain an inoculum (a mycoviral transcript from an artificial infectious clone or a plasmid) for fungal protoplasts transfection or transformation; to isolate high-quality *B. cinerea* protoplasts; to perform fungal protoplasts inoculation with the obtained inoculum; and to do protoplasts fusion assays between infected and non-infected *B. cinerea* strains for mycovirus transfer.

2 Materials

Sterilize all media and material in an autoclave at 120°C for 20 min, unless otherwise specified.

2.1 Media

1. Potato dextrose agar (PDA) plates: dissolve 39 g of premixed PDA in 1 L of distilled water or mix potato dextrose broth with 2% agar and add 1 L of distilled water. Alternatively, mix 15 g of bacteriological agar, 20 g of dextrose, and 4 g of potato starch in 1 L of distilled water. Pour 20 mL of medium in 90 mm diameter plates at no more than 60°C.
2. Potato dextrose broth (PDB): dissolve 26.5 g of premixed Potato dextrose broth in 1 L of distilled water. Alternatively, mix 20 g of dextrose and 6.5 g of infusion from potatoes in 1 L of distilled water.
3. Osmostabilized SH agar: 0.6 M sucrose, 5 mM HEPES pH 5.3, 1 mM $(NH_4)_2HPO_4$, 1.2% agar. To prepare SH agar medium

2×, add 1.2 g of HEPES to 400 mL of bi-distilled water. Adjust to pH 5.3. Then add 12 g of agar. Add 500 mL of bi-distilled water. Keep at 60°C after autoclaving. Mix with 500 mL of 1.6 M sucrose to obtain the 1× SH agar medium.

4. Sucrose 1.6 M: Add 273.84 g of sucrose to a final volume of 500 mL of bi-distilled water. Sterilize by filtering or autoclaving at 110°C.
5. Luria-Bertani (LB) medium: dissolve 10 g Bacto-tryptone, 5 g yeast extract, and 5 g NaCl in 1 L of distilled water. Adjust to pH 6.8–7.2. Dispense in 0.5 L bottles and autoclave.

2.2 Protoplast Generation (see Note 1)

6. Miracloth.
7. Funnel.
8. Filter paper.
9. Cell counting chamber.
10. KC buffer: 0.6 M KCl, 50 mM $CaCl_2$, pH 6. Dissolve 8.95 g of KCl and 1.48 g of $CaCl_2$ $2H_2O$ in 180 mL of bidistilled water and adjust to pH 6 with KOH. Fill up to 200 mL and sterilize by filtering. Store at 4°C.
11. Syringe and 0.45 μm syringe-driven filters.
12. Vinotaste Pro (Novozymes), a mix of fungal cell wall- degrading enzymes.
13. Conical centrifuge tubes.
14. Petri dishes.
15. Bi-distilled water.
16. Drigalski spatula.
17. 250 mL Erlenmeyer flask.
18. 100 mL Erlenmeyer flask.
19. Pasteur pipette or cut tips.
20. Water bath with shaker incorporated.
21. Rotary shaker.

2.3 Preparation of Nucleic Acid for Transfection

22. Bacterial culture tubes.
23. Resuspension solution: Add 1.26 mL of 50% glucose solution, 1.4 mL of EDTA 0.5 M and 1.75 mL of Tris-HCl pH 8 1 M. Fill up to a volume of 70 mL. Keep at 4°C.
24. Lysis solution: Prepare fresh. Mix 8.8 mL of bi-distilled H_2O, 0.2 mL of NaOH 10 N and 1 mL of SDS 10%. Keep at RT.
25. Neutralization solution: Add 60 mL of potassium acetate 5 M and 11.5 mL of glacial acetic acid. Fill up to a volume of 100 mL. Keep at 4°C.
26. 1.5 mL microcentrifuge tubes.

27. Isopropanol.
28. 75% ethanol.
29. Nuclease-free bi-distilled water.
30. Restriction enzyme and its buffer.
31. Nucleic-acid loading buffer: Add 3.9 mL of glycerol, 500 μL of 10% SDS, 200 μL of 0.5 M EDTA, 0.025 g of bromophenol blue, and 0.025 g of xylene cyanol. Bring to 10 mL total volume with nuclease-free bi-distilled H_2O.
32. Agarose and TAE 1× buffer for electrophoresis: for a 1% gel, add 1 g of agarose to 100 mL of TAE 1× buffer and boil them until the agarose is completely dissolved. Then, pour it on the gel casting tray and insert the comb to create wells for sample loading. Allow the agarose gel to solidify for approximately 30 min.
33. TAE 50×: Dissolve 242 g of Tris-base in 700 mL of bi-distilled water, add 57.1 mL of 100% glacial acetic acid, and 100 mL of 0.5 M EDTA pH 8.0. Bring it up to the final volume of 1 L.
34. Electrophoresis equipment and power supply.
35. PCR purification kit.
36. MEGAscript™ T7 Transcription Kit (Invitrogen) or T7 RNA polymerase.
37. Thermocycler.

2.4 Transfection

38. 25% PEG solution: 25% PEG-3350, 50 mM $CaCl_2$, 10 mM Tris-HCl pH 7.5. Add 2.5 g of PEG-3350, either 0.5 mL of 1 M $CaCl_2$ or 0.04 g of $CaCl_2 \cdot 2H_2O$, and 0.1 mL of 1 M Tris-HCl pH 7.5 to a final volume of 10 mL. Sterilize by filtering. Keep at RT.
39. 1 M Tris-HCl pH 7.5: Dissolve 121.1 g Tris in 800 mL of bi-distilled water. Add approximately 65 mL concentrated HCl to bring pH to 7.5. Wait until the solution reaches room temperature (RT), then measure the pH to confirm. Adjust volume to 1 L.
40. 1.5 mL microcentrifuge tubes.
41. Spermidine.
42. Antibiotic for selection.

2.5 Protoplast Fusion

43. STE buffer: 1 M sorbitol, 100 mM $CaCl_2$, 100 mM Tris-HCl pH 8. Dissolve 36.43 g of sorbitol and 2.96 g of $CaCl_2 \cdot 2H_2O$ in 180 mL of bi-distilled water. Add 0.1 mL of 1 M Tris-HCl pH 8 and bring volume to 200 mL. Store at 4°C.

44. 1 M Tris-HCl pH 8: Dissolve 121.1 g Tris in 800 mL of bi-distilled water. Add approximately 42 mL concentrated HCl to bring pH to 7.5. Wait until the solution reaches RT, then measure the pH to confirm. Adjust volume to 1 L.
45. 1 M sorbitol: Dissolve 36.43 g of sorbitol in 180 mL of bi-distilled water and bring the volume to 200 mL.

3 Methods

3.1 Preparation of *Botrytis cinerea* Protoplasts

Conduct the next procedures under a laminar hood or under the flame to maintain sterility. The protocol is graphically detailed in Fig. 1.

1. Grow *B. cinerea* on PDA Petri dishes for 3–7 days at 24°C (*see* **Note 2**).
2. Inoculate a 250 mL Erlenmeyer flask containing 100 mL of PDB medium with 10^6 conidia. You can collect conidia by adding 2 mL of sterile bi-distilled water to the plates. Scratch

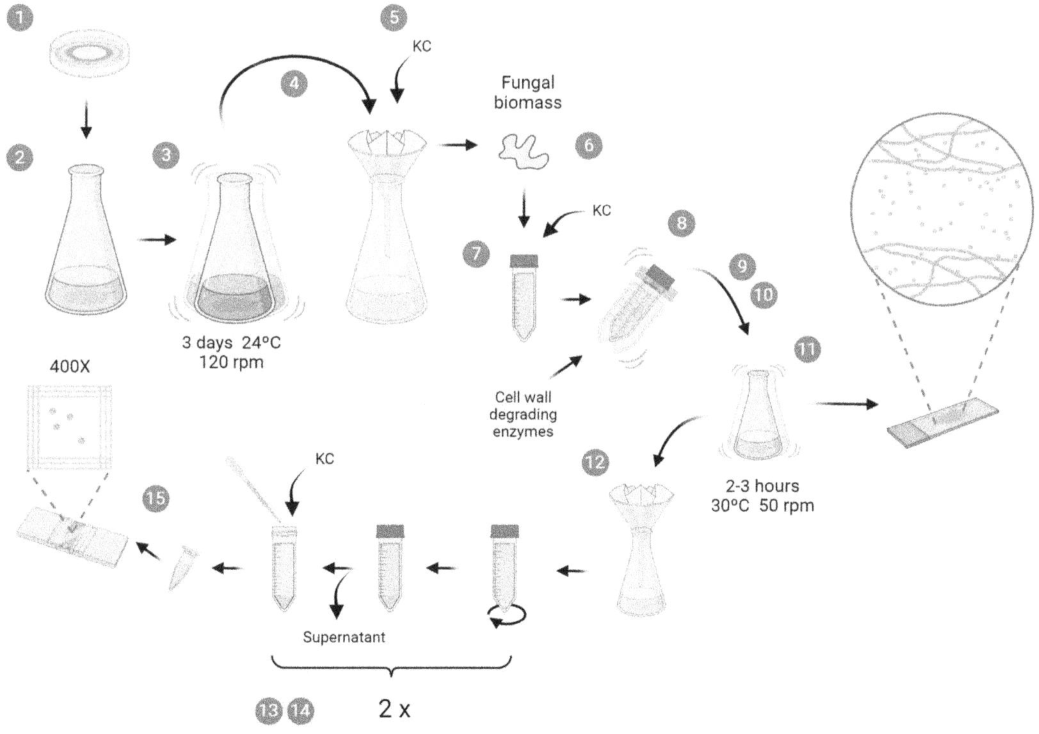

Fig. 1 Graphical representation of the preparation of *B. cinerea* protoplasts. The numbers correspond to the steps detailed in Sect. 3.1

the mycelia gently using a previously sterilized Drigalski spatula to free the conidia from the hyphae. Collect the conidia and hyphae suspension and pass them through a sterile Miracloth filter (*see* **Note 3**). Quantify the filtered conidia suspension using a cell counting chamber. Alternatively, take some fungal biomass and add it directly to the PDB.

3. Grow the culture for 3 days at 24°C and 120 rpm on a rotary shaker.
4. Filter the fungal biomass by passing the culture through Miracloth using a funnel.
5. Wash the mycelia twice with sterile bi-distilled water and once with KC buffer.
6. Collect the mycelia on the Miracloth and dry them using sterile filter paper (*see* **Note 4**).
7. Transfer the mycelia to a sterile 50 mL conical centrifuge tube. Fill it with up to 50 mL of KC buffer and resuspend the mycelia using a vortex.
8. Add 200 mg of Vinotaste and vigorously shake until achieving homogenization.
9. Incubate for 5 min on ice.
10. Shake the digestion again and transfer it to a 100 mL Erlenmeyer flask.
11. Incubate the digestion for 2–3 hin a water bath at 30°C and 50 rpm (*see* **Note 5**).
12. Filter the digestion mixture through Miracloth to get rid of mycelial debris. Collect the filtered protoplast suspension in a 50 mL conical centrifuge tube. Wet the filter with cold KC buffer before adding the protoplast solution (*see* **Note 6**).
13. Fill the conical tube to 50 mL with cold KC buffer. Pellet the protoplasts by centrifugation at 2000 × *g* for 10 min.
14. Discard the supernatant and resuspend the protoplasts in cold KC buffer carefully using a Pasteur pipette. Centrifuge again at 2000 × *g* for 10 min.
15. Discard the supernatant and resuspend the protoplasts in 1 mL of cold KC buffer. Quantify the protoplasts using a cell counting chamber (*see* **Note 7**).

3.2 Preparation of Inoculum for Protoplast Inoculation

Depending on the approach, it may be useful to create a selectable *B. cinerea* strain, either by expressing a fluorescent protein or an antifungal resistance gene, which can serve as a tool for protoplast fusion or transfection. Therefore, in this section we describe first the generation of genetic material for transformation, but the same protocol can be applied for the generation of the starting material for the *in vitro* transcription from a mycovirus cDNA infectious clone. The protocols are graphically detailed in Figs. 2 and 3.

3.2.1 For Plasmid DNA (see Note 8)

1. Inoculate LB medium supplemented with the appropriate antibiotic with *Escherichia coli* DH5α carrying the plasmid containing the infectious clone (*see* **Note 9**).
2. Incubate the culture at 37°C and 150 rpm overnight.
3. Fill a 1.5 mL microcentrifuge tube with the bacterial culture and centrifuge it for 1 min at 13,000 rpm.
4. Discard the supernatant, add 200 μL of resuspension buffer, and vigorously resuspend it using a vortex or pipette.
5. Add 200 μL of lysis solution and carefully mix by inverting the tube three or four times (*see* **Note 10**).
6. Add 200 μL of neutralizing solution and mix carefully by inversion. Cell debris will become visible (*see* **Note 11**).
7. Centrifuge the tubes at 13,000 rpm for 15 min.
8. Transfer the supernatant to a new microcentrifuge tube with the pipette, carefully avoiding the pellet (*see* **Note 12**).
9. Add 500 μL of isopropanol and mix by inversion.
10. Incubate the tubes for 15 min at –20°C.

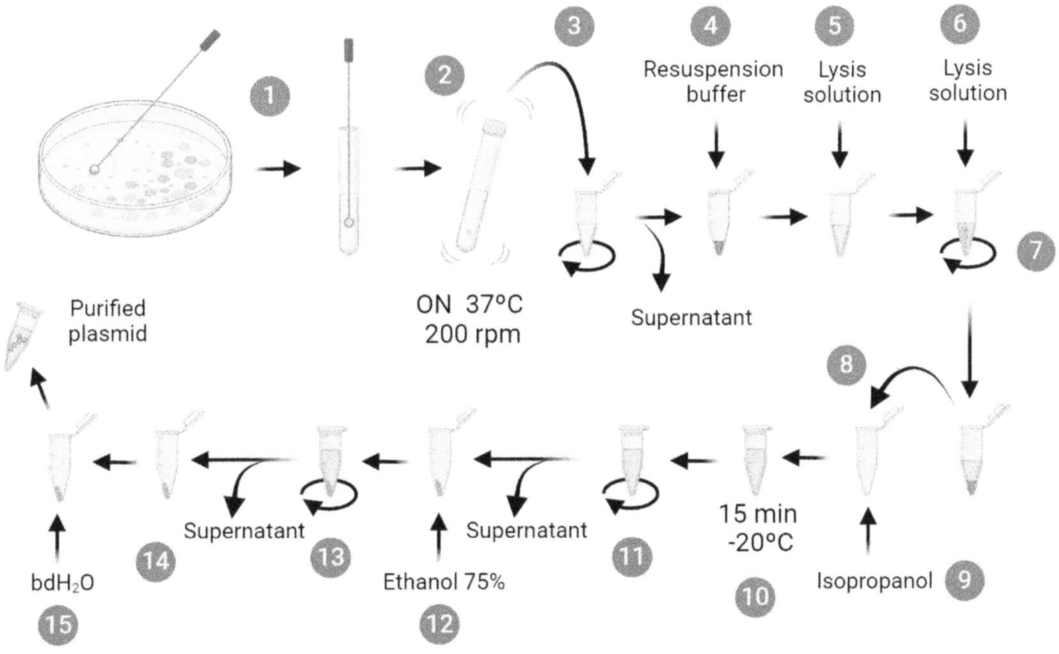

Fig. 2 Graphical representation of the purification of plasmid DNA from *E. coli*. The numbers correspond to the steps detailed in Sect. 3.2.1

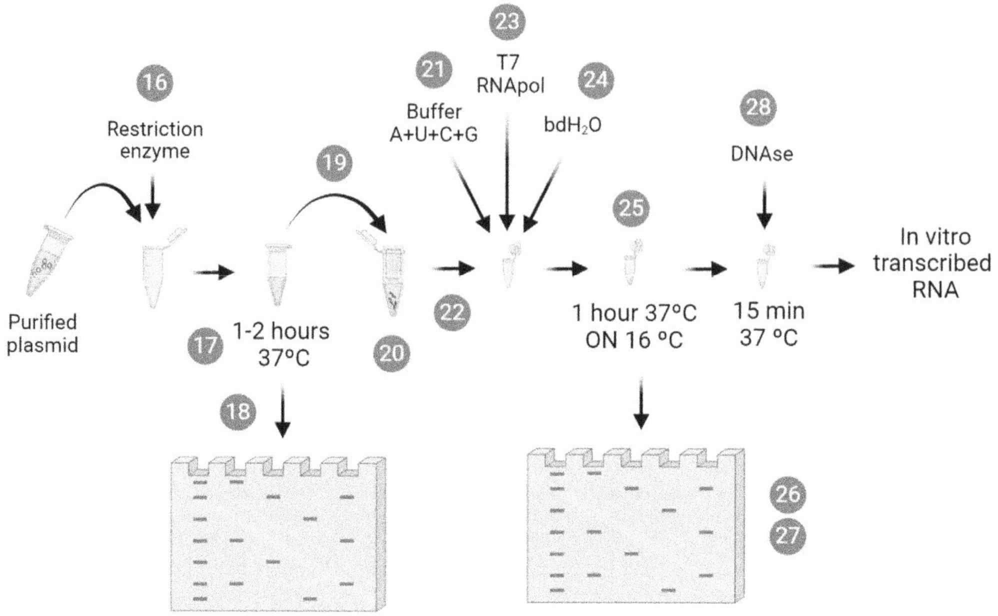

Fig. 3 Graphical representation of the *in vitro* transcription. The numbers correspond to the steps detailed in Sect. 3.2.2

11. Centrifuge the tubes for 10 min at 13,000 rpm.
12. Discard the supernatant, add 250 μL of 75% ethanol, and mix it with a vortex.
13. Centrifuge the tubes for 5 min at 7500 rpm.
14. Discard the supernatant and let the DNA pellet air-dry (*see* **Note 13**).
15. Resuspend the pellets in nuclease-free bi-distilled water and quantify the plasmid DNA using a NanoDrop.
16. Use the plasmid DNA to transform *B. cinerea* protoplasts.

*3.2.2 For Positive Single-Stranded RNA Mycovirus Infectious Clones (*see *Note 14)*

17. To serve as a template for the *in vitro* transcription reaction, prepare a digestion with 5 μg of the previously obtained plasmid using one unit of the appropriate restriction enzyme and the corresponding buffer.
18. Incubate the reaction for 1–2 h or overnight at 37°C.
19. Check the efficiency of the digestion by electrophoresis. Load a small aliquot of the reaction with DNA loading buffer on a 1% agarose gel. Run the electrophoresis at 100 V for 45 min.
20. If the plasmid has been fully digested, purify the reaction using an appropriate gel and PCR purification kit following the manufacturer's instructions.

21. Elute the DNA in nuclease-free bi-distilled H_2O and quantify it using a nanodrop (*see* **Note 15**).
22. Prepare the *in vitro* transcription reaction following the manufacturer's instructions. For MEGAscript™ T7 Transcription Kit (Invitrogen™), in a volume of 20 μL, add 2 μL of 10× reaction buffer, 2 μL of ATP solution, 2 μL of CTP solution, 2 μL of UTP solution, and 2 μL of a 1:5 dilution of the GTP solution (*see* **Note 16**) in a 200 μL Eppendorf tube.
23. Add 1 μg of the linearized template DNA (*see* **Note 17**).
24. Add 2 μL of T7 enzyme and mix.
25. Complete the reaction with up to 20 μL of nuclease-free bi-distilled water.
26. Incubate the reaction for 1 h at 37°C and then at 16°C overnight in a thermocycler (*see* **Note 18**).
27. Visualize the result of the *in vitro* transcription reaction by loading 0.25 μL mixed with RNA loading buffer in a 1% agarose gel prepared with nuclease-free bi-distilled water.
28. Run the electrophoresis at 75 V for 1 h and check the gel under UV light.
29. While the gel is running, add 1 μL of Turbo-DNase to the reaction and incubate it for 15 min at 37°C.
30. Use the *in vitro* transcribed RNA (*see* **Note 19**) to transfect *B. cinerea* protoplasts.

3.3 Inoculation of *Botrytis cinerea* Protoplasts

The protocol for protoplast inoculation is graphically detailed in Fig. 4.

1. Adjust the concentration to 10^8 protoplasts per mL and prepare 100 μL aliquots of the suspension in 1.5 mL sterile microcentrifuge tubes (*see* ***Note 20***).
2. Add 100 μL of 25% PEG solution to each tube.
3. Mix 5 μL of spermidine 50 mM with 40 μL of the nucleic acid of interest for fungal transformation or transfection (*see* **Note 21**). Use water for the control. Transfer it to the protoplast containing microcentrifuge tubes.
4. Incubate the mixture for 20 min on ice.
5. Add 500 μL of the 25% PEG solution to each tube and mix it gently by inversion or tapping.
6. Incubate the mixture for 10 min at RT.
7. Add 500 μL of cold KC buffer.
8. Mix the protoplasts with 50 mL of SH agar medium at 50°C–55°C, split the mixture, and spread it in two Petri dishes. Cool to RT and incubate at 24°C (*see* **Note 22**).

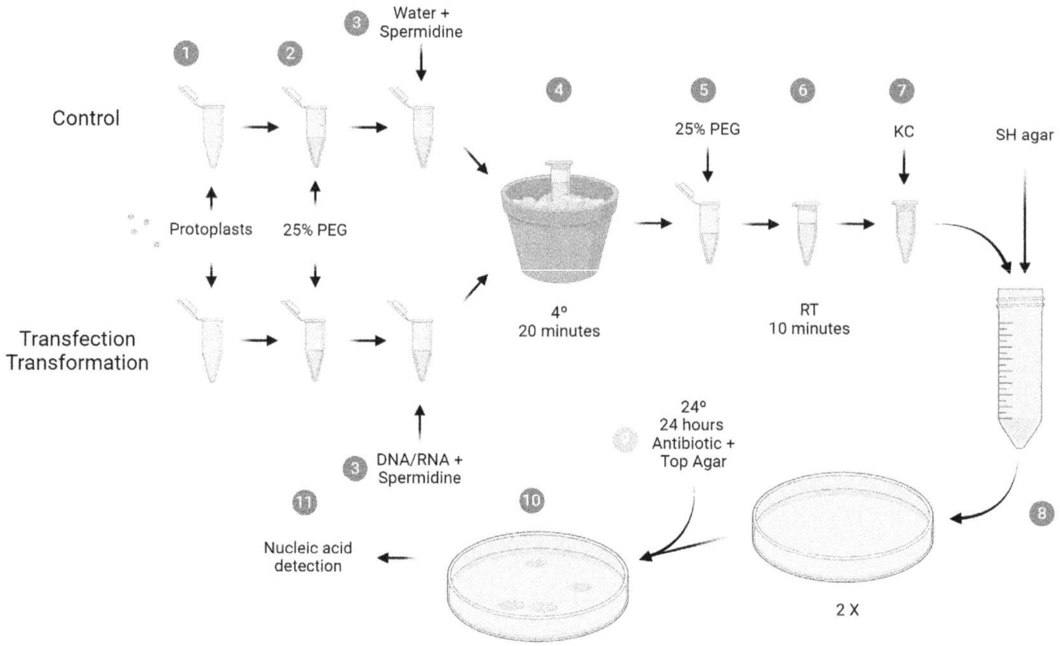

Fig. 4 Graphical representation of *B. cinerea* protoplasts inoculation. The numbers correspond to the steps detailed in Sect. 3.3. Step 9, marked in light grey, as indicated in the text, is optional depending on the experimental strategy used

9. If the introduced nucleic acid contains an antibiotic selection marker, such as hygromycin, after 24 h overlay the Petri dish with 5 mL of top agar containing appropriate antibiotic.
10. Wait till colonies appear in the case of selective media or until the mycelia cover the Petri dish in the case of no selection.
11. Check for the presence of the nucleic acid (plasmid or mycovirus) with an appropriate technique (*see* **Note 23**).

3.4 Generation of a Fungal Isolate from a Single Protoplast Infected with a Mycovirus

The protocol for generation of a fungal isolate derived from a single protoplast is graphically detailed in Fig. 5.

1. Prepare and quantify protoplasts as previously stated.
2. Set a final concentration of 2×10^2 protoplasts/mL of KC buffer (*see* **Notes 24 and 25**).
3. Place 100 μL of this final solution in the center of a Petri dish with 20 mL of SH regeneration medium (*see* **Note 26**).
4. Spread the volume over the Petri dish, with the help of a Drigalski spatula (*see* **Note 27**).
5. Incubate for 2–3 days at 24°C.
6. Transfer the independent colonies of *B. cinerea* to separate PDA Petri dishes with the help of a scalpel or a yellow tip.

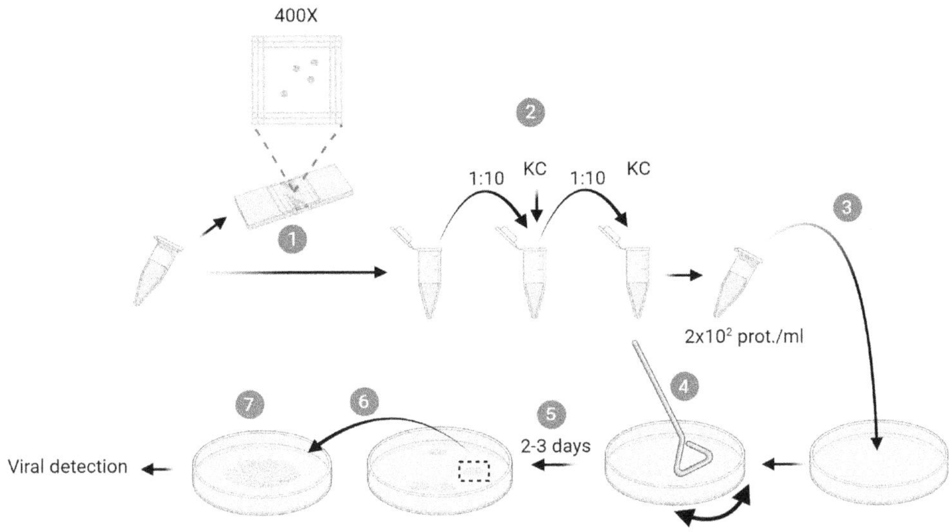

Fig. 5 Graphical representation of the generation of independent *B. cinerea* colonies from a single protoplast infected with a mycovirus. The numbers correspond to the steps detailed in Sect. 3.4

7. Let the isolates grow and check for the presence of the mycovirus with an appropriate technique (*see* **Note 23**).

3.5 Protoplast Fusion for Mycovirus Transference

The protocol for mycovirus transference through protoplast fusion is graphically detailed in Fig. 6.

1. Select the donor and receptor *B. cinerea* strains for mycoviral transference and prepare and quantify protoplasts as previously stated.
2. Pellet the protoplast solution by centrifugation at 2000× *g* for 6 min.
3. Gently resuspend the pellet in 4:1 STE/25% PEG solution to achieve a final concentration of 3×10^7 protoplasts/mL (see **Note 28**).
4. Take 100 μL of the protoplast suspension and dilute it with 500 μL of ice-cooled 1 M sorbitol.
5. Spin down cells in a microcentrifuge tube at 2000× *g* for 6 min at 4°C. Remove supernatant gently by aspiration with the 1 mL pipette.
6. Resuspend the pellet in 500 μL of cold 1 M sorbitol.
7. Mix the two types of protoplasts gently.
8. Incubate on ice for 30 min.
9. Add 1 mL of PEG solution and mix gently by inversion.
10. Incubate the mixture for 25 min at RT.

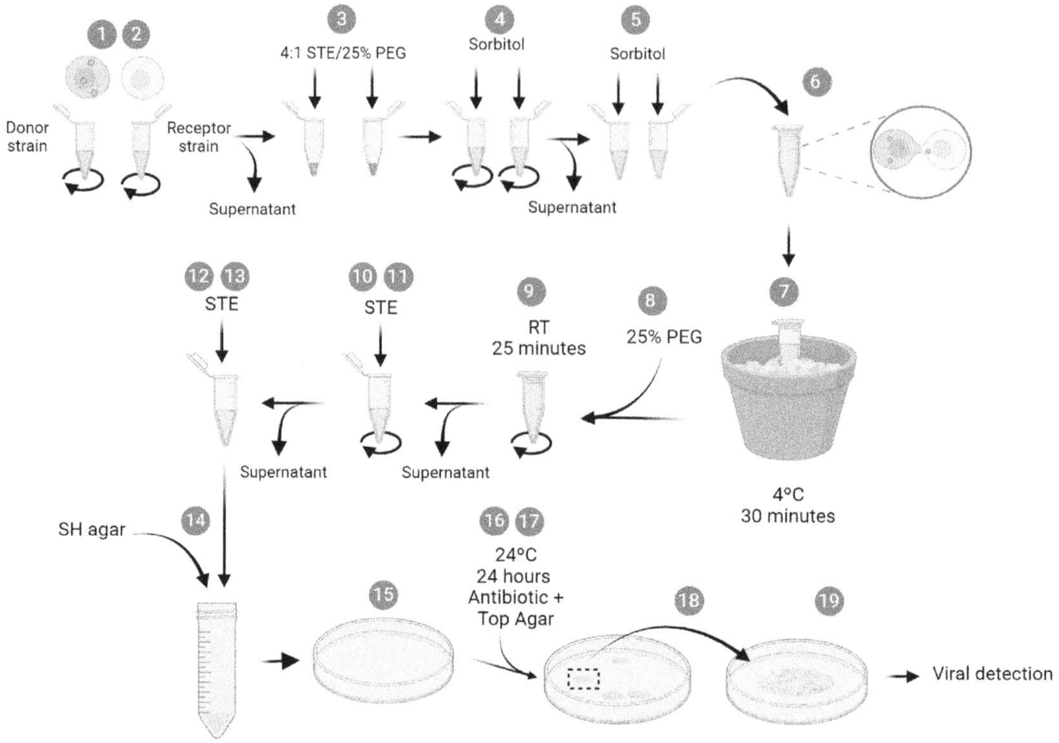

Fig. 6 Graphical representation of the *B. cinerea* protoplasts fusion protocol for mycovirus transference. The numbers correspond to the steps detailed in Sect. 3.5

11. Pellet the protoplasts by centrifugation at 2000 × *g* for 6 min at 4°C and remove the supernatant by aspiration.
12. Add 1 mL of cool STE and mix gently.
13. Pellet the protoplasts at 2000 × g for 1 min at 4°C and remove the supernatant by aspiration.
14. Add 500 μL of cool STE to the tube and mix gently until the pellet is dissolved.
15. Mix 125 μL of the fused protoplast suspension with 20 mL of SH medium (46°C–48°C) and pour it into a Petri dish (*see* **Note 16**).
16. Let it cool to RT and incubate at 24°C overnight.
17. If the receptor fungal strain carries a selection marker, add 5 mL of top agar containing the appropriate antibiotic for selection.
18. Incubate for 2–3 days at 24°C.
19. Transfer the independent colonies of *B. cinerea* to separate PDA Petri dishes with the help of a scalpel or a yellow tip.

20. Let the isolates grow and check for the presence of the mycovirus with an appropriate technique (*see* **Note 23**).

4 Notes

1. Protoplasts are cells that have lost their cell wall, becoming quite sensitive to osmotic stress, so all buffers used during their generation must be osmostabilized.
2. Cultivating the fungus under cycles of light and darkness will promote conidiation.
3. If Miracloth is not available, you can use a sterile borosilicate filter (porosity 0) or blue 1 mL tips filled with medical cotton.
4. Fold the Miracloth with the mycelia inside and wrap it with sterile filter paper; use as many as necessary. Apply pressure to extract the moisture.
5. Check periodically the formation of protoplasts during the cell wall degradation process to both assess the appearance of protoplasts and their physical qualities. Protoplasts should look round and turgid under the microscope. Protoplasts should start appearing after 30–60 min of incubation, so it is advisable to take 10 μL aliquots every 45–60 min of digestion for examination.
6. Extending the incubation period excessively could potentially compromise the viability or competency of the protoplasts. It is recommended to filter them within a maximum of 2–3 h after their initial appearance.
7. To prevent regeneration of the cell wall, keep protoplasts on ice.
8. We describe here a homemade protocol for the extraction of bacterial plasmid DNA, but similar results or even higher purity can be obtained with a commercial extraction kit.
9. The antibiotic for selection will depend directly on the plasmid in which it is encoded.
10. When bacteria are lysed, the mix will go from cloudy to clear, and it will become thicker and stickier.
11. Overly aggressive mixing of the solutions will cause cellular debris to break up, increasing the amount of impurities and bacterial chromosomal DNA during subsequent DNA precipitation. After addition of neutralizing solution, start inverting the tube and when no more cell debris appears, proceed to the next step.

12. Carry-over of cellular debris and chromosomal DNA will decrease the final concentration of plasmid, decreasing both the efficiency of the transfection and the *in vitro* transcription.
13. Incomplete drying of the pellet will cause traces of ethanol to remain in the sample, which may reduce efficiency or even prevent its use in subsequent enzymatic reactions.
14. For steps involving RNA, always use certified nuclease-free pipette tips and tubes, and wear gloves. RNase contamination can seriously compromise the integrity of RNA samples. Keep RNA at 4°C unless otherwise specified to avoid degradation.
15. To increase the amount of purified linear plasmid, you can heat the double-distilled H_20 to 60°C, collect the sample, and put it again on the same elution column of the kit.
16. The *in vitro* transcription reactions outlined here are prepared in 20 μL, but they can be performed in a smaller volume, depending on the kit used.
17. Before adding the DNA, make sure that you have a sufficiently high concentration of linearized plasmid to perform the experiment. DNA can be concentrated by precipitating and resuspending it in a smaller volume or using a Vacuum Concentrator System.
18. The yield and quality of the resulting transcripts significantly improve when the incubation is extended overnight in our experience, but reactions can be performed in less time at 37°C.
19. Keep a small volume of the *in vitro* transcription reaction and dilute it to be used as a positive control for mycoviral detection after the transfection.
20. When preparing the aliquots, a control tube will be needed that will be mixed just with the 25% PEG solution, the spermidine, and just water, instead of the nucleic acid.
21. A standard fungal transfection with DNA can be done with between 2 and 10 μg of a DNA mycovirus infectious clone. A transfection with an RNA mycovirus infectious clone is the result of *in vitro* transcription and is normally not quantified. Use [9, 25] as references.
22. Control the temperature of the SH agar; maintaining the SH agar 2× at 55°C could positively impact protoplast viability. The hot SH 2× agar can be mixed with RT 1.6 M sucrose just before adding the protoplasts to prevent excessively high temperatures. Spread the protoplasts promptly afterward to prevent the medium from solidifying.
23. The method for detection of the transfected mycovirus will greatly depend on the nature of its genetic material and its

ability to accumulate inside the fungal cell. DNA extraction and PCR detection, RNA extraction followed by RT-PCR or other methods can be applied. You can find more information in [9].

24. The optimal concentration of the working protoplast solution could be variable depending on the results obtained. Regeneration rate of protoplasts could vary because of the *B. cinerea* protoplast state and the strain used, so higher concentrations could be used when regeneration is compromised. However, using higher concentrations would make it more difficult to differentiate independent colonies.
25. An appropriate manner to achieve an accurate dilution of the protoplasts solution would include serial dilutions, using KC buffer to maintain omostability. Paying attention to correct dilution and quantification of the protoplasts will allow setting an appropriate final concentration, placing only a few protoplasts per plate. After protoplast regeneration, this would allow us to transfer independent colonies in the final step of the protocol.
26. The number of replicates must also be considered to obtain the desired number of independent colonies. Set as many replicates as needed.
27. Both the media and the protoplasts solution need to be poured homogeneously in all plates, avoiding irregularities on the surface, so that individual single-protoplast colonies can be picked up.
28. The preparation described in the first step can be made as follows: Centrifuge at 2000 × *g* for 6 min at 4°C a solution of protoplasts with known volume and concentration. Resuspend it with four parts of STE and one part of PEG solution in the needed total volume to obtain a 3×10^7 concentration.

Acknowledgments

J.P.-M. is supported by a postdoctoral fellowship, Margarita Salas MSALAS-2022-20257, from Universidad de Sevilla, financed by the European Union–NextGenerationEU funds through Ministerio de Universidades. J.M.-G. is financially supported by the Project I+D+i PID2020-120106RB-I00, supported by MCIN/AEI/10.13039/501100011033. L.C. was supported by two postdoctoral contracts from the European Union's Horizon 2020 Research and Innovation Program (Grant Agreement No. 773567) and from the "Severo Ochoa Program for Centers of Excellence in R&D" (2017–2021) of the Agencia Estatal de Investigación of

Spain (Grant SEV-2016-0672 to CBGP). This study was financially supported by VIROPLANT, a project that received funding from the European Union's Horizon 2020 Research and Innovation Program (Grant Agreement No. 773567), and by the Project I+D+i PID2020-120106RB-I00, supported by MCIN/AEI/10.13039/501100011033.

References

1. Li P, Wang S, Zhang L et al (2020) A tripartite ssDNA mycovirus from a plant pathogenic fungus is infectious as cloned DNA and purified virions. Sci Adv 6:eaay9634. https://doi.org/10.1126/sciadv.aay9634
2. Feng C, Feng J, Wang Z et al (2021) Identification of the viral determinant of hypovirulence and host range in Sclerotiniaceae of a genomovirus reconstructed from the plant metagenome. J Virol 95:e00264-21. https://doi.org/10.1128/JVI.00264-21
3. Chen B, Choi GH, Nuss DL (1994) Attenuation of fungal virulence by synthetic infectious hypovirus transcripts. Science 264:1762–1764. https://doi.org/10.1126/science.8209256
4. Esteban R, Fujimura T (2003) Launching the yeast 23S RNA narnavirus shows 5′ and 3′ cis-acting signals for replication. Proc Natl Acad Sci USA 100:2568–2573. https://doi.org/10.1073/pnas.0530167100
5. Moleleki N, van Heerden SW, Wingfield MJ et al (2003) Transfection of *Diaporthe perjuncta* with *Diaporthe* RNA virus. Appl Environ Microbiol 69:3952–3956. https://doi.org/10.1128/AEM.69.7.3952-3956.2003
6. Marzano S-YL, Hobbs HA, Nelson BD et al (2015) Transfection of *Sclerotinia sclerotiorum* with in vitro transcripts of a naturally occurring interspecific recombinant of *Sclerotinia sclerotiorum* hypovirus 2 significantly reduces virulence of the fungus. J Virol 89:5060–5071. https://doi.org/10.1128/JVI.03199-14
7. Zhang R, Hisano S, Tani A et al (2016) A capsidless ssRNA virus hosted by an unrelated dsRNA virus. Nat Microbiol 1:15001. https://doi.org/10.1038/nmicrobiol.2015.1
8. Wang Q, Mu F, Xie J et al (2020) A single ssRNA segment encoding RdRp is sufficient for replication, infection, and transmission of ourmia-like virus in fungi. Front Microbiol 11:379. https://doi.org/10.3389/fmicb.2020.00379
9. Córdoba L, Ruiz-Padilla A, Rodríguez-Romero J, Ayllón MA (2022) Construction and characterization of a Botrytis virus F infectious clone. J Fungi 8:459. https://doi.org/10.3390/jof8050459
10. Córdoba L, Ruiz-Padilla A, Pardo-Medina J, Rodríguez-Romero JL, Ayllón MA (2024) Construction of a Mycoviral Infectious Clone for Reverse Genetics in Botrytis cinerea. Methods Mol Biol 2751:47–68. https://doi.org/10.1007/978-1-0716-3617-6_4
11. Rodríguez-García C, Medina V, Alonso A, Ayllón MA (2014) Mycoviruses of *Botrytis cinerea* isolates from different hosts. Ann Appl Biol 164:46–61. https://doi.org/10.1111/aab.12073
12. Donaire L, Rozas J, Ayllón MA (2016) Molecular characterization of *Botrytis ourmia*-like virus, a mycovirus close to the plant pathogenic genus *Ourmiavirus*. Virology 489:158–164. https://doi.org/10.1016/j.virol.2015.11.027
13. Howitt RLJ, Beever RE, Pearson MN, Forster RLS (2006) Genome characterization of a flexuous rod-shaped mycovirus, Botrytis virus X, reveals high amino acid identity to genes from plant 'potex-like' viruses. Arch Virol 151:563–579. https://doi.org/10.1007/s00705-005-0621-y
14. Donaire L, Pagán I, Ayllón MA (2016) Characterization of *Botrytis cinerea* negative-stranded RNA virus 1, a new mycovirus related to plant viruses, and a reconstruction of host pattern evolution in negative-sense ssRNA viruses. Virology 499:212–218. https://doi.org/10.1016/j.virol.2016.09.017
15. Ruiz-Padilla A, Rodríguez-Romero J, Gómez-Cid I, Pacifico D, Ayllón MA (2021) Novel mycoviruses discovered in the mycovirome of a necrotrophic fungus. mBio 12:e03705–e03720
16. Ruiz-Padilla A, Turina MAM (2023) Molecular characterization of a tetra segmented ssDNA virus infecting *Botrytis cinerea* worldwide. Virol J 20:306
17. Hao F, Zhou Z, Wu M, Li G (2017) Molecular characterization of a novel endornavirus from the phytopathogenic fungus *Botrytis cinerea*. Arch Virol 162:313–316. https://doi.org/10.1007/s00705-016-3106-2
18. Hao F, Wu M, Li G (2021) Characterization of a novel genomovirus in the phytopathogenic fungus Botrytis cinerea. Virology 553:111–116. https://doi.org/10.1016/j.virol.2020.11.007
19. Hao F, Wu M, Li G (2018) Molecular characterization and geographic distribution of a mymonavirus in the population of *Botrytis*

cinerea. Viruses 10:432. https://doi.org/10.3390/v10080432
20. Hao F, Ding T, Wu M et al (2018) Two novel hypovirulence-associated mycoviruses in the phytopathogenic fungus *Botrytis cinerea*: molecular characterization and suppression of infection cushion formation. Viruses 10:254. https://doi.org/10.3390/v10050254
21. Khalifa ME, MacDiarmid RM (2021) A mechanically transmitted DNA mycovirus is targeted by the defence machinery of its host, *Botrytis cinerea*. Viruses 13:1315. https://doi.org/10.3390/v13071315
22. Howitt RLJ, Beever RE, Pearson MN, Forster RLS (2001) Genome characterization of Botrytis virus F, a flexuous rod-shaped mycovirus resembling plant 'potex-like' viruses. J Gen Virol 82:67–78. https://doi.org/10.1099/0022-1317-82-1-67
23. Donaire L, Ayllón MA (2017) Deep sequencing of mycovirus-derived small RNAs from *Botrytis* species. Mol Plant Pathol 18:1127–1137. https://doi.org/10.1111/mpp.12466
24. Svanella-Dumas L, Marais A, Faure C et al (2018) Genome characterization of a divergent isolate of the mycovirus Botrytis virus F from a grapevine metagenome. Arch Virol 163:3181–3183. https://doi.org/10.1007/s00705-018-3975-7
25. Zhang L, Wang S, Ruan S et al (2023) A mycovirus VIGS vector confers hypovirulence to a plant pathogenic fungus to control wheat FHB. Adv Sci 10:e2302606. https://doi.org/10.1002/advs.202302606

Check for updates

Chapter 13

Assessing Sensitivity to Antifungal Plant Compounds or Agrochemicals

H. M. Suraj, André Fleissner and Jan A. L. van Kan

Abstract

Botrytis cinerea, a major fungal pathogen of crops, has been extensively studied for its resistance to antifungal plant metabolites and agrochemicals. Evaluating the sensitivity of *B. cinerea* to these compounds is crucial for understanding its pathogenicity and for developing effective plant protection strategies. This study describes two methods for antifungal sensitivity assessment: the spot assay, which visualizes fungal growth on agar medium supplemented with antifungal compounds, and the resazurin assay, which quantifies fungal metabolic activity in liquid culture. Using saponins as antifungal compounds, we demonstrate the utility of these methods on *B. cinerea*. These methods provide simple, reliable, and reproducible protocols to assess the sensitivity of antifungal compounds on *B. cinerea*.

Keywords Antifungal sensitivity assays, Plant-derived saponins, Spot assay, Resazurin assay, Fungal resistance mechanisms

1 Introduction

Understanding the mechanisms of fungal sensitivity to antifungal compounds is critical for addressing the growing challenge of antifungal resistance, which poses a significant threat to agriculture and food security. This chapter provides a guide to two such methods: spot assay and resazurin assay. The spot assay allows for the visual assessment of fungal growth on solid media supplemented with antifungal compounds, while the resazurin assay quantitatively measures fungal metabolic activity in liquid culture using a fluorescence-based approach. Together, these assays provide complementary insights into fungal behavior under antifungal stress. In this chapter, we applied these methods to evaluate the sensitivity of two *B. cinerea* strains—M3a, a saponin-sensitive strain, and various transformants designed to test specific mechanisms of antifungal resistance. The transformants include strains modified to detoxify saponins or protect against them via membrane repair. These protocols described

Luca Nerva and Walter Chitarra (eds.), *Botrytis cinerea*, Methods in Molecular Biology, vol. 3076,
https://doi.org/10.1007/978-1-0716-5551-1_13,

below are simple yet robust to assess the sensitivity of *B. cinerea* to antifungal compounds.

2 Materials

2.1 For Spot assay

2.1.1 Media

Liquid minimal medium (MM) for microscopy is prepared using the following recipe [1]:

Component	Quantity
Sucrose	20 g
Vogel's solution	20 mL
pH (no need to adjust; it results in 5.8–6)	
H_2O	1000 mL

2.1.2 Stock Solutions

Vogels Stock Solutions (50×)

Component	Quantity
Sodium citrate	125 g
KH_2PO_4	250 g
NH_4NO_3	100 g
$MgSO_4 \times 7H_2O$	10 g
$CaCl_2 \times 2H_2O$	5 g
Chloroform (optional, only to keep sterile)	5 mL
Trace elements stock solution	5 mL
Biotin stock (50 mg in 1 L of 50% EtOH)	2.5 mL
ddH_2O	755 mL

Trace Elements Stock Solutions

Component	Quantity
Citric acid	5 g
$ZnSO_4 \times 7H_2O$	5 g
$Fe(NH_4)_2(SO_4)_2 \times 6H_2O$	1 g
$CuSO_4 \times 5H_2O$	0.25 g
$MnSO_4 \times H_2O$	0.05 g
Boric acid	0.05 g
$Na_2MoO_4 \times 2H_2O$	0.05 g
ddH_2O	100 mL

BDES (20× Stock)

Component	Quantity
L-Sorbose	20 g
D-Fructose	1 g
Sucrose	1 g
ddH_2O	100 mL

All the above-mentioned stock solutions need to be autoclaved and stored at 4°C.

2.1.3 Stock Solution of Antifungal Compounds

Based on the polarity of the antifungal compound, decide what solvent is required to dissolve the compound to make a stock solution. Here, we consider three antifungal plant metabolites to test antifungal activity.

1. α-Tomatine (20 mg/mL stock): 20 mg of α-tomatine (T3636, TCI Europe) is dissolved in 75% MeOH containing 0.1% formic acid.
2. Digitonin (20 mg/mL stock): 20 mg of digitonin (Carl Roth, HN76.3) is dissolved in 1 mL ddH_20 and mixed well by vortexing.
3. α-Hederin (20 mg/mL stock): 20 mg of α-hederin (Carl Roth) is dissolved in 1 mL DMSO and mixed by vortexing.

2.1.4 BDES Agar Medium

Component	Quantity
Vogels stock solutions (50×)	2 mL
Sucrose	2 g
Agar	1.5 g
pH (no need to adjust; it results in 5.8–6)	
ddH_20	95 mL
Autoclave	
BDES stock (20×) autoclaved	5 mL

2.2 Resazurin Assay

2.2.1 Media

RPMI with Glucose and Resazurin Liquid medium prepared as follows:

Component	Quantity
RPMI 1640 (R8755; Sigma)	2.08 g
MOPS	6.9 g
D-Glucose	4 g

Component	Quantity
Milli-Q water	Adjust to 100 mL

RPMI + Glucose (2× Solution)

i. Dissolve all components (RPMI 1640, MOPS, and D-glucose) in approximately 80 mL of Milli-Q water with stirring.
ii. Adjust the pH to 7.0 using concentrated NaOH (e.g., 5 M).
iii. Adjust the final volume to 100 mL with Milli-Q water.
iv. Sterilize the solution by filtration. Store the solution in the refrigerator until use.

Resazurin Stock Solution (10×, 0.02% w/v)

Component	Quantity
Resazurin sodium salt (R7017; Sigma)	20 mg
Milli-Q water	Adjust to 100 mL

i. Dissolve resazurin in Milli-Q water.
ii. Sterilize the solution by filtration.
iii. Store the solution at room temperature in a dark container.

Working Medium Preparation

Freshly prepare 1× RPMI medium supplemented with 0.002% resazurin immediately before use as follows:

i. Combine the following:
   a. 25 mL of 2× RPMI + glucose solution.
   b. 5 mL of 10× resazurin stock solution.
   c. 20 mL of sterile Milli-Q water.
ii. Mix thoroughly.
iii. Use within a couple of hours (discard any leftover medium).

3 Methods

3.1 Spot Assay

A spot assay is a method used for evaluating the growth of microorganisms, such as fungi, under specific conditions. It has long been applied to yeast species [2] but is also suitable for filamentous fungi, provided they produce spores. In this technique, small volumes (typically 5–10 μL) of serially diluted spore suspensions, ranging from higher to lower concentrations, are carefully spotted onto an agar medium supplemented, for example, with an antifungal compound. The grid-based layout is used for the simultaneous testing of multiple strains on the same plate, facilitating direct comparison between the strains. After incubation, growth patterns, such as colony number and growth density, are assessed to determine the effects of the antifungal compound. Unlike radial growth assays, the spot assay employs a medium with high osmotic potential, which restricts fungal colonies from expanding radially. As a result, the colony size remains small, enabling a semi-quantitative assessment of fungal growth under antifungal stress. By using a dilution range, the spot assay is particularly suitable for observing subtle differences that would otherwise not be detected in radial growth assays.

Case study: In this demonstration, we applied the spot assay to study the wild-type (WT) strain *B. cinerea* M3a, which is sensitive to saponins, a class of plant-derived antifungal compounds. The WT strain was compared to various transformants generated in previous work [3]. These transformants include:

1. Strains capable of detoxifying saponins:
   a) M3a + BcTom1 (Tomatinase from B05.10)
   b) M3a + Cftom1 (Tomatinase from *Cladosporium fulvum*).
2. Strain that protects itself against saponins through membrane repair
   a) M3a + BcGT28a (Sterol glycosyl transferase family 28)

3.1.1 Preparation of BDES Agar Medium with Antifungal Compound

a) Autoclave Vogel's Minimal Medium (MM) with agar and the BDES stock solution (20×) separately.
b) After autoclaving, mix the agar medium and the BDES stock to achieve a final concentration of 1× BDES agar.
c) Allow the medium to cool to approximately 55°C.
d) Add the antifungal stock solution to the medium at three different concentrations (low, medium, and high). Prepare two replicate plates for each antifungal concentration.

1. **Preparation of spore suspension:** In a laminar airflow cabinet, prepare a 1 mL spore suspension stock for different *B. cinerea* strains at a concentration of 2×10^7 spores/mL in 1.5 mL Eppendorf tubes. Prepare serial dilutions (1:10) of the stock to create suspensions with concentrations of:

- 2×10^6 spores/mL
- 2×10^5 spores/mL
- 2×10^4 spores/mL

2. **Spotting the spores on BDES agar medium with antifungal compound:**

a) Mark each agar plate with a 4 × 4 grid (four rows and four columns) for spotting. Each row will be inoculated with a different *B. cinerea* strain, and each column will have a different spore concentration (highest to lowest) as shown in Fig. 1.

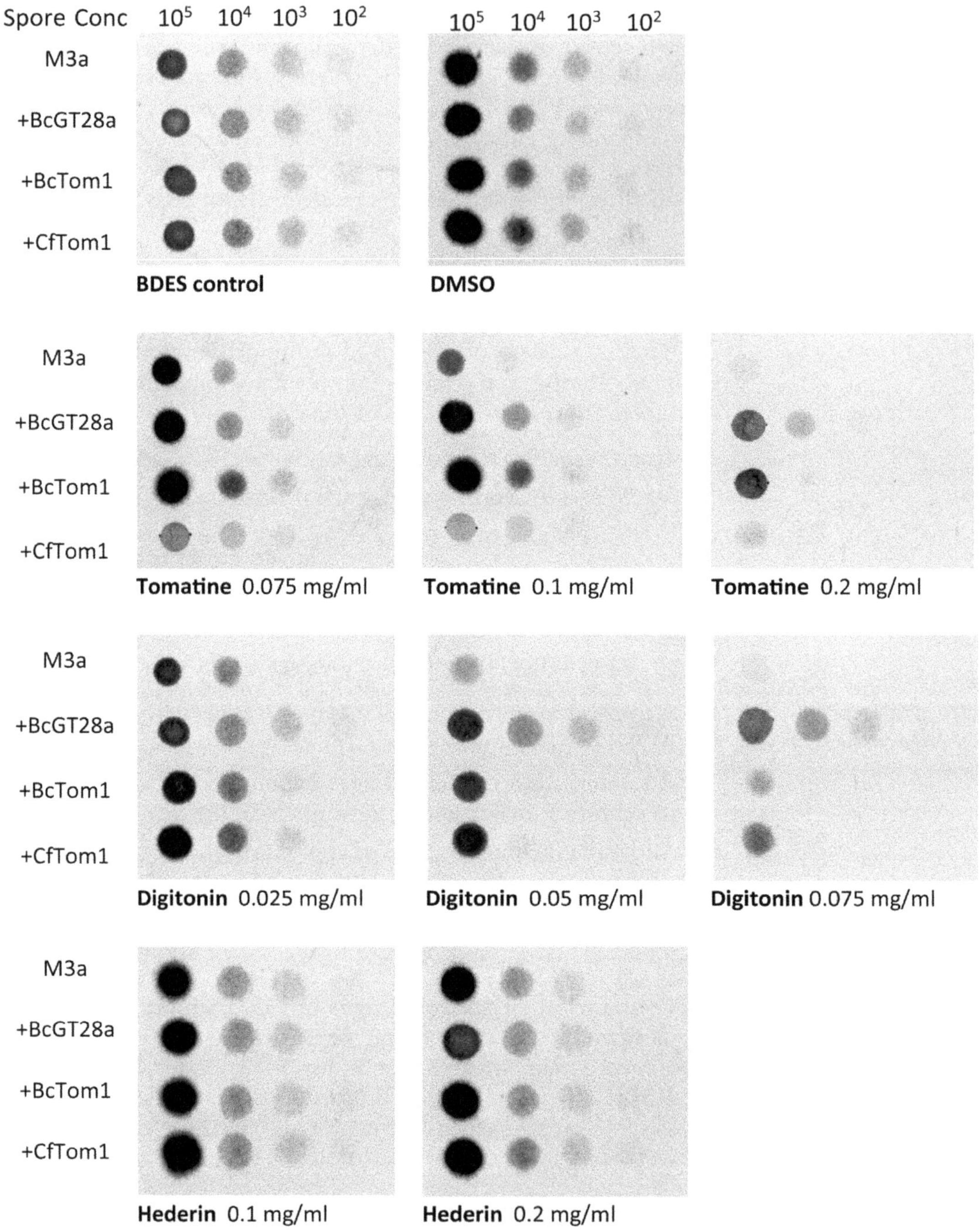

Fig. 1 Spot assay of *Botrytis cinerea* M3a (WT) and transformants on saponin-containing BDES agar medium. This figure shows the growth of *Botrytis cinerea* M3a and three separate transformants on BDES agar medium supplemented with varying concentrations of saponins (α-tomatine, digitonin, and α-hederin). Spores were spotted as serial dilutions from higher to lower concentrations (left to right) for each strain, ranging from 10^5 to 10^2 spores per droplet. DMSO was used as a control, since it served as the solvent for the compounds. Images were taken 3 days post-inoculation (3 DPI). The growth patterns demonstrate the differential tolerance of the strains to saponins. BcGT28a (membrane repair protein) provides better protection against saponins than tomatinase in the spot assay

b) Vortex the 2×10^7 spores/mL suspension thoroughly, then pipette 5 μL onto the first spot in the grid. This will result in 10^5 spores per spot.

- Similarly, vortex and pipette:

  2×10^6 spores/mL onto the second spot in the grid, which results in 10^4 spores/spot.

  2×10^5 spores/mL onto the third spot, which results in 10^3 spores/spot.

  2×10^4 spores/mL onto the fourth spot, which results in 10^2 spores/spot.

- Repeat this process for each strain in the designated rows.
- Allow the droplets to air-dry before taking the plates out of the clean bench.
- Incubate the plates at 20°C for 3 days.

4 Note

- Accurate counting of spores is crucial for reproducibility, so it is necessary to vortex the spore suspension well before counting.
- Allowing the droplets to air-dry before removing the plates from the clean bench is essential to ensure that each inoculation spot has the same diameter and to prevent the spore suspension from spreading during movement.
- Any droplet that is spilled elsewhere on the plate may form a colony, which may confound the interpretation.

4.1 Resazurin Assay

In this method, resazurin is used to measure the metabolic activity of fungi. Metabolically active fungal cells reduce the non-fluorescent dye resazurin to the fluorescent dye resorufin. This fluorescence is measured, and the intensity is proportional to the number of metabolically active cells. This assay has been widely used to test antifungal susceptibility for various microorganisms, most notably for determining MICs (minimum inhibitory concentrations) for the human pathogenic fungus *Aspergillus fumigatus* [4]. Here, we describe this simple and sensitive protocol using two strains of *B. cinerea*, M3a and M3aΔ*pef1*. BcPEF1 is a membrane repair protein, and the knock-out of BcPEF1 in M3a makes it more sensitive to α-tomatine [3]. We used these two strains with α-tomatine as an antifungal to describe the resazurin assay.

4.1.1 Prepare the Medium

Freshly prepare the 1× RPMI medium with resazurin as follows:

- Prepare a 2× RPMI + glucose stock and 10× resazurin stock.

- Mix 25 mL of 2× RPMI + glucose with 5 mL of 10× resazurin and 20 mL of sterile Milli-Q water. Use the medium immediately for the assay.

4.1.2 Prepare the 96-Well Plate

- **Add RPMI medium:**
  - Pipette 100 μL of 1× RPMI medium with resazurin into each well.
- **Add antifungal compound:**
  - Add 50 μL of the antifungal compound (4× concentrated) to each well.
  - Use a gradient of concentrations starting from 20 μg/mL and decreasing incrementally (e.g., 17.5 μg/mL, 15 μg/mL, to 0 μg/mL).
  - Perform three technical replicates for each concentration.
- **Add spore suspension:**
  - Add 50 μL of spore suspension (final concentration: 1×10^5 spores/mL, 4× concentrated). Also use three wells as blanks (refer to Fig. 2), where you do not add spore suspension but add 50 μL of 1× RPMI medium.
  - Ensure the layout follows the gradient pattern shown in Fig. 2.

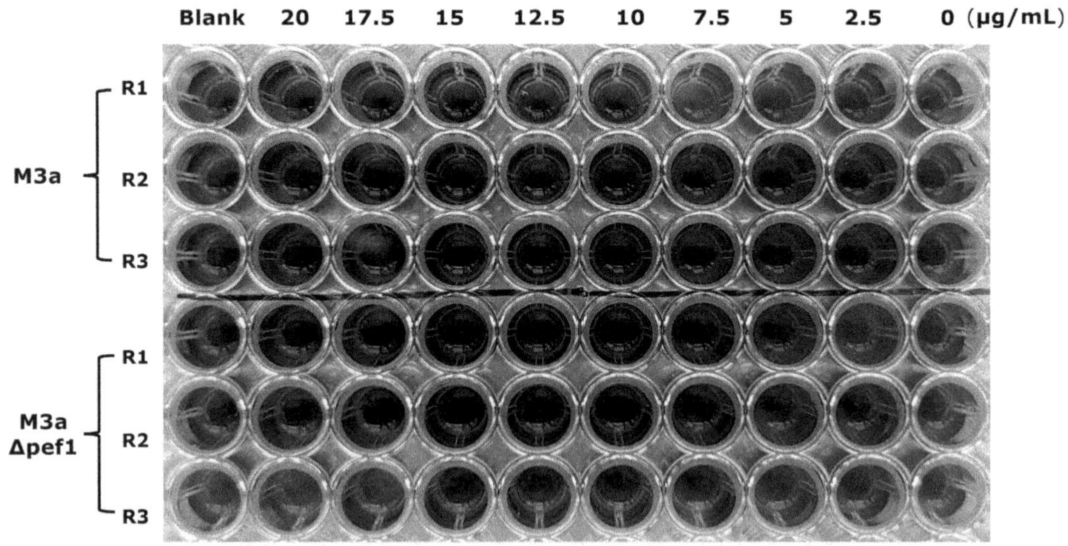

Fig. 2 Layout of 96-well plate for resazurin assay. The plate layout has a gradient of antifungal concentrations from 20 μg/mL to 0 μg/mL in triplicate for the two fungal strains (*B. cinerea* M3a and M3aΔ*pef1*). Wells containing RPMI medium without spores are considered blanks. Pink color represents higher metabolic activity, while blue/purple color indicates reduced fungal growth with low or no metabolic activity

4.1.3 Incubation

i. Cover the plate with aluminum foil.
ii. Incubate the plate in the dark at 20°C for 24 h.

4.1.4 Measurement and Calculating IC90 Value

i) **Measure fluorescence:**
Use a plate reader to measure fluorescence with excitation at 570 nm and emission at 615 nm.

ii) **Data analysis:**

- Subtract the raw fluorescence values of the blanks from each sample.
- Calculate the average fluorescence for the three replicates for each concentration.
- Normalize the data to the "0 μg/mL" control to calculate the percentage growth.

iii) **Plot the curve:**

- Plot normalized growth (%) against antifungal concentration. Refer to Fig. 3 for an example of the plotted growth inhibition curves plotted for the *B. cinerea* strains M3a vs. M3aΔ*pef1* sensitivity to α-tomatine.
- Add a horizontal line at 10% growth to determine the IC90 value (the lowest concentration where growth is inhibited below 10%).

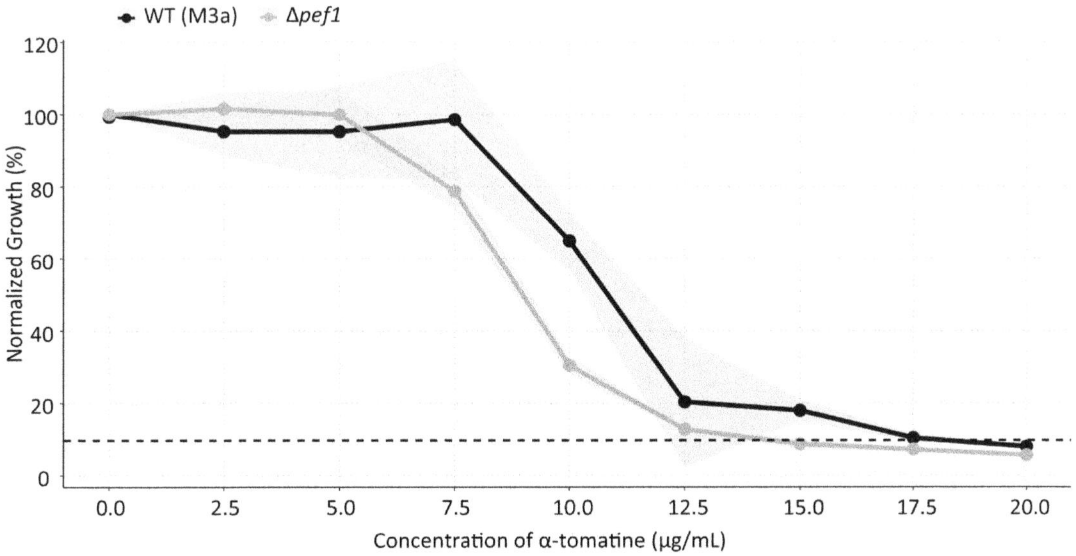

Fig. 3 This graph depicts the normalized growth (%) of two *B. cinerea* strains, M3a (yellow line) and M3aΔ*pef1* (orange line), in response to varying α-tomatine concentrations (20–0 μg/mL). Fluorescence measurements were adjusted for blank wells, and growth was normalized to the 0 μg/mL α-tomatine. The black dashed line represents the 10% growth threshold, indicating the IC90. M3aΔ*pef1* exhibits greater sensitivity to α-tomatine, with growth dropping below the threshold at lower concentrations compared to the WT M3a

5 Note

- Freshly prepare 1× RPMI medium supplemented with resazurin, as the dye degrades over time.
- Include blanks (no spores) and controls (no antifungal compound) for each replicate.
- Do not put fungal samples in the outer rows and columns of the 96-well plates, as these can lead to inconsistent results due to evaporation from the edges of the plate. Leave the outer rows and columns empty or fill with sterile water.
- Follow a clear gradient of antifungal concentrations for accurate IC90 determination.
- Cover plates with aluminum foil to minimize light-induced degradation of resazurin.

Acknowledgments

H.M.S. acknowledges the TKI Graduate School Green Top Sectors (grant LWV21.291, TU-2021-20), The Netherlands, for funding his project. We thank Dr. Martin Weichert (Laboratory of Genetics, Wageningen University) for helping to optimize the protocol for the resazurin assay.

References

1. Vogel HJ (1956) A convenient growth medium for *Neurospora crassa*. Microb Genet Bull 13:42–47
2. Sahu SR, Utkalaja BG, Patel SK, Acharya N (2023) Spot assay and colony forming unit (CFU) analyses-based sensitivity test for *Candida albicans* and *Saccharomyces cerevisiae*. Bio Protoc 13(21):e4872. https://doi.org/10.21769/BioProtoc.4872
3. Monteiro MC, De La Cruz M, Cantizani J, Moreno C, Tormo JR, Mellado E, De Lucas JR, Asensio F, Valiante V, Brakhage AA, Latgé JP, Genilloud O, Vicente F (2012) A new approach to drug discovery: high-throughput screening of microbial natural extracts against *Aspergillus fumigatus* using resazurin. J Biomol Screen 17(4):542–549. https://doi.org/10.1177/1087057111433459
4. You Y, Suraj HM, Matz L, Herrera Valderrama AL, Ruigrok P, Shi-Kunne X, Pieterse FPJ, Oostlander A, Beenen HG, Chavarro-Carrero EA, Qin S, Verstappen FWA, Kappers IF, Fleißner A, van Kan JAL (2024) Botrytis cinerea combines four molecular strategies to tolerate membrane-permeating plant compounds and to increase virulence. Nat Commun 15(1), 6448. https://doi.org/10.1038/s41467-024-50748-5

Chapter 14

Extracellular Vesicle Isolation from *Botrytis cinerea* Culture and *B. cinerea*-Infected *Arabidopsis* Plants

Lida Halilovic, Simoné Murguia, Yuan Wang, Huaitong Wu and Hailing Jin

Abstract

Extracellular vesicles (EVs) are universally conserved and essential players in the communication between cells across nearly all organisms of life. *Botrytis cinerea* produces and secretes EVs that are critical in its virulence on plants. Numerous studies in various fungal species reveal that EVs are important across different pathosystems to facilitate fungal infection. For characterization of the cargo and role of EVs in fungal growth and communication, it is necessary to carefully and precisely purify fungal EVs from fungal culture or the plant interaction interface. Here, we present a method for the isolation of *Botrytis cinerea* EVs from culture and the infection site of *B. cinerea* on *Arabidopsis thaliana*. Fungal EVs isolated using this method are free from cellular contamination from fungal or plant cells, remain viable in functional assays, and provide high-quality RNA and protein.

Keywords Extracellular vesicles, Fungal virulence, Plant–microbe interactions

1 Introduction

Extracellular vesicles (EVs) are nanoscale membrane-bound particles that carry biological materials such as DNA, RNA, proteins, and metabolites. They serve as a critical form of cellular communication within organisms and between interacting organisms and are produced by cells from all kingdoms of life. EV research conducted studying intra-organismal molecular transport in mammalian cells has revealed unique and impactful cargo with distinct physiological effects, highlighting the importance of EVs in cellular interactions. Fungal EVs have been observed in over 20 species of yeast and filamentous fungi [1]. Multiple modes of biogenesis are known for fungal EVs, including being derived from multivesicular bodies or budding from the plasma membrane [2, 3]. Numerous plant-pathogenic fungal species utilize EVs to deliver effectors during colonization of host plants [4]. As well, EVs play a significant role in

Luca Nerva and Walter Chitarra (eds.), *Botrytis cinerea*, Methods in Molecular Biology, vol. 3076,
https://doi.org/10.1007/978-1-0716-5551-1_14,

the virulence of human fungal pathogens and the modulation of mammalian host immune responses to fungal infection [5, 6].

Botrytis cinerea produces and secretes EVs containing dynamic and bioactive cargoes that are critical for its interactions with other organisms [7–9]. During plant infection, *B. cinerea* utilizes EVs to traffic sRNA effectors into plant cells for cross-kingdom gene silencing [7]. Numerous homologs of mammalian EV protein markers have been putatively identified in *B. cinerea* EVs, and the tetraspanin protein Punchless 1 (*Bc*PLS1) has been validated as a biomarker of *B. cinerea* EVs [7]. *Bc*PLS1 is essential for fungal pathogenicity, emphasizing the critical role of fungal EVs in the infection process. *B. cinerea* EVs have been shown to be internalized into interacting fungal cells such as *Fusarium graminearum* and into host plant cells through clathrin-mediated endocytosis [7, 8]. EVs have been isolated from further plant pathogenic fungal species, either purified from fungi grown in culture or directly from infected plant tissue. These include but are not limited to EVs from *Ustilago maydis*, *Penicillium digitatum*, multiple *Fusarium* species, *Blumeria hordei*, *Zymoseptoria tritici*, and *Colletotrichum higginsianum* [4]. Fungal EVs from various pathogens contain diverse cargoes of nucleic acids, proteins, metabolites, and polysaccharides.

This method details the isolation of *B. cinerea* EVs from fungal culture and the interface of *B. cinerea*-infected *Arabidopsis* plants. Samples are processed using differential centrifugation and filtering to remove larger debris. Then, either ultracentrifugation or the EXODUS H-600 machine can be used to obtain final concentrated EV pellets [10]. EVs can be further purified by discontinuous sucrose density gradient centrifugation. These methods herein can easily be tailored to isolate EVs from other fungal species or the disease interface of other fungal-infected plants.

2 Materials

2.1 Fungal Culturing

1. *Botrytis cinerea* strain B05.10.
2. Potato dextrose agar plates.
3. Purified YEPD: 0.3% yeast extract, 1% peptone, 2% glucose, pH 6.5 medium, ultracentrifuged and filtered through a 0.22 μm flask filter.
4. Double-distilled sterile water.
5. 70 μm cell strainer.
6. 50 mL conical tubes.
7. 0.22 μm flask filters.
8. Tabletop rocker.
9. −80°C freezer.

2.2 B. cinerea Inoculation

1. *Botrytis cinerea* strain B05.10.
2. *Arabidopsis thaliana* wild type (ecotype Col-0).
3. Growth chamber.
4. 70 μm cell strainer.
5. Fungal inoculation buffer: 1% Sabouraud maltose broth buffer in water.
6. Sprayer (Preval®, Chicago Aerosol, USA).

2.3 Fungal Extracellular Vesicle Purification from Culture or Infected Plants

1. Phosphate-buffered saline (PBS): 137 mM NaCl, 2.7 mM KCl, 10 mM Na_2HPO_4, 1.8 mM KH_2PO_4 filtered through 0.22 μm flask filter.
2. Infiltration buffer: 20 mM MES, 2 mM $CaCl_2$, 0.1 M NaCl, pH 6.0.
3. 10%–90% sucrose solutions made with DEPC-treated water.
4. 50 mL centrifuge tubes.
5. 1 mL needless syringes.
6. 0.45 μm syringe filters.
7. Oak Ridge tubes.
8. Clear tape.
9. Oakridge tubes.
10. Tabletop centrifuge with a swinging-bucket rotor of appropriate size for 50 mL tubes and a fixed-angle rotor of appropriate size for Oak Ridge tubes.
11. Ultracentrifuge compatible with a SW32.1 Ti rotor.

3 Methods

3.1 Botrytis cinerea Culturing

1. Store glycerol stocks of *B. cinerea* in a −80°C freezer (*see* **Note 1**).
2. Inoculate a potato dextrose agar plate with 5–10 μL of *B. cinerea* glycerol stock conidia suspension. Allow it to grow for 10 days at room temperature, without sealing the plate so air can flow to allow sporulation.
3. Prepare liquid culture media by centrifuging YEPD at 100,000 × *g* for 1 h using an ultracentrifuge and swing-bucket rotor (SW32.1 Ti) to eliminate any particles, then filter the media through a 0.22 μm filter prior to use.
4. Harvest *B. cinerea* conidia from a 10-day-old plate by adding 1 mL of sterile water onto the plate, then use a bacterial loop to liberate conidia into the water. Pipette up the mycelia and conidia suspension and filter it through a 70 μm cell strainer

to remove mycelia. Using a hemocytometer, calculate the concentration of the collected conidia.

5. For one EV sample, prepare 100 mL of purified YEPD and inoculate with *B. cinerea* conidia to obtain a final amount of 10^5 total spores in 100 mL.
6. Culture the *B. cinerea* for 48 h at room temperature by placing the culture on a rocker with gentle rocking (*see* **Note 2**).

3.2 *B. cinerea* EV Isolation Using Ultracentrifugation

1. Collect culture supernatant by centrifuging in 50 mL conical tubes at 3000 × *g* for 15 min at 4°C to pellet mycelia (*see* **Note 3**).
2. Filter the supernatant through a 70 μm cell strainer to further remove cells.
3. Centrifuge the cell-free supernatant in Oak Ridge tubes at 10,000 × *g* for 30 min to remove smaller debris, then collect the supernatant in a new tube and centrifuge again at 10,000 × *g* for 30 min to ensure the removal of aggregates.
4. Filter the resulting supernatant through a 0.45 μm filter (*see* **Note 4**).
5. In 38 mL ultracentrifuge tubes, ultracentrifuge the processed supernatant at 100,000 × *g* for 1 h in a swing-bucket rotor (SW32.1 Ti) to pellet EVs.
6. Remove the ultracentrifuge supernatant and gently resuspend the pellet in 100 μL filtered PBS, then add 25 mL of filtered PBS to fill the ultracentrifuge tube.
7. Ultracentrifuge the sample again at 100,000 × *g* for 1 h, remove the supernatant, and resuspend the final pellet in 100 μL filtered PBS. Store EVs at −80°C or use fresh (*see* **Note 5**).

3.3 *B. cinerea* EV Isolation Using EXODUS H-600

1. Process the fungal culture the same as in steps 1–4 from Sect. 3.2.
2. Pour the fungal culture into a sterile container, either a 15 mL or 50 mL Falcon tube or the 250 mL EXODUS-provided sample tube, and keep the sample on ice (*see* **Note 6**).
3. Fill the "Buffer" bottle with autoclaved and filtered PBS (store the tube at 4°C when the machine is off) and fill the "Cleaner" tube with 1% bleach diluted with autoclaved ultrapure water.
4. The main screen will show an error when the machine is turned on: "Daily cleaning cycle." Select the error and press "Resolve."
5. The machine will open a hatch door in which the EID labeled "Cleaning EID" can be placed. Insert the EID with the QR code on the left. Once the EID is in place, select "OK"

on the main screen. This will commence the 30 min cleaning process.

6. Once the cleaning is done, the hatch will open again for the removal of the Cleaning EID. Remove it and press "EID In."
7. Turn on the UV light to sterilize the EXODUS sample chamber (30 min).
8. Press the "EID Out" button on the EXODUS machine main screen and place the large cartridge with 70 nm size cutoff in the chamber (LA.A08) (*see* **Note 7**).
9. Press "EID In." Place the fungal sample in the sample chamber.
10. Press "Start" on the main screen. The machine will scan the QR code on the EID and then prompt the user to input the sample processing information, including the sample name, sample volume, and program to use. For fungal EVs, use the "Cell Culture Medium" program and the "Strong" setting. Press the "Done" button, confirm the run information, and the run will begin.
11. The estimated completion time will show in red. Monitor the machine for errors that may arise, including low volumes of buffer, cleaner, or water, or the request to empty the waste container (*see* **Note 8**).
12. When the run is complete, the run information will be stored in the machine log for future reference. Press the "EID Out" button.
13. The packaging with the EID will have a small plastic bag with two rubber plugs. Before removing the EID, place the rubber plugs on either side of the EID to cover the openings. These plugs prevent the sample from being lost during resuspension.
14. Remove the EID and hold it upright at chest level so you can see down the EID into the filter area. Pipet 1 mL of autoclaved and filtered PBS and eject the buffer through the filter area, sweeping the pipet tip along the length of the filter area to wash all filter surfaces. The buffer will run down the filter and collect on one side of the EID. Tilt the EID to the collection side and take up the buffer, repeating the elution of the EVs from the filter for a total of 20 washes.
15. After the elution, take up the resuspended EVs and store them in a 1.5 mL tube. Store the EVs at −80°C or use fresh (*see* **Note 9**).

3.4 Isolation of Fungal EVs from B. cinerea-Infected Arabidopsis

1. Grow 144 *Arabidopsis thaliana* plants until they are 3–4 weeks old after transplanting, at the stage of growth where they are mature but have not begun to flower.

2. Infect the plants by harvesting conidia from 10-day-old *B. cinerea* plates as described in steps 1, 2, and 4 in Sect. 3.1. Dilute the conidia in 40 mL of fungal inoculation buffer to a final concentration of 10^4 spores/mL. Use an aerosolizing sprayer to spray the 40 mL of the conidia suspension equally across 144 plants. Store the infected plants in an airtight container that has a thin layer of water along the bottom. Allow the infection to progress overnight for 16 h (*see* **Note 10**).
3. Harvest plant leaves by cutting off the leaves at the base, excluding the petiole, and wash the leaves twice in beakers of distilled water. Briefly dry the leaves on a paper towel.
4. Place approximately 30 mL (by volume) of leaves in a 100 mL needleless syringe and fill the syringe to 50 mL with infiltration buffer. Replace the plunger of the syringe and push it in to eliminate the air in the syringe.
5. Cover the tip of the syringe with a folded piece of Parafilm and slowly pull back on the plunger, then slowly and gently release the plunger three times to infiltrate the buffer into the leaves (*see* **Note 11**).
6. Remove the leaves from the syringe onto a paper towel. Work in batches to infiltrate all leaves, because infiltrated leaves must be processed quickly to avoid cellular damage.
7. Place a two-foot-long strip of clear tape down and adhere a 1 mL needleless syringe to one end of the tape. Blot the excess liquid off the infiltrated leaves, then carefully place each leaf side by side, slightly overlapping, on the tape parallel to the syringe with the cut sides oriented downward along the length of the 1 mL syringe (Fig. 1) (*see* **Note 12**).
8. Roll the syringe across the leaves, rolling them into a lollipop shape, then place another piece of tape across the top of the 1 mL syringe to secure the roll (*see* **Note 13**).
9. Place the roll into a 50 mL conical tube and centrifuge in a fixed-angle rotor at 900 × *g* for 10 min. All centrifugations are done at 4°C.
10. Remove the rolls from the tubes. The liquid in the base of the tube is the apoplastic wash fluid (AWF). Pool the apoplastic wash fluid together into one conical tube.
11. Centrifuge the AWF at 2000 × *g* for 30 min in a swing-bucket rotor.
12. Filter the supernatant through a 0.45 μm syringe filter, then centrifuge in an Oak Ridge tube at 10,000 × *g* for 30 min in a fixed-angle rotor.
13. Transfer the supernatant to an ultracentrifuge tube and ultracentrifuge at 100,000 × *g* for 1 h. Remove the supernatant

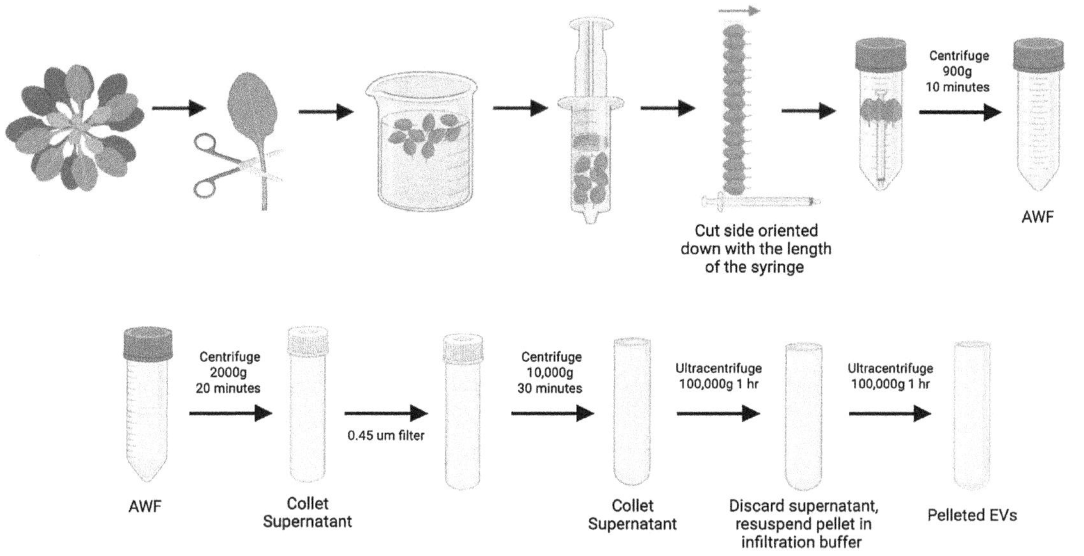

Fig. 1 Isolation of EVs from *B. cinerea*-infected *Arabidopsis thaliana.* Leaves are harvested and infiltrated with infiltration buffer. Infiltrated leaves are arranged on tape and rolled around a syringe, then centrifuged to collect apoplastic wash fluid (AWF). AWF is processed by differential centrifugation to eliminate cell debris and larger particles. Cleaned AWF is ultracentrifuged to pellet EVs. Created in https://BioRender.com

and resuspend the pellet in 100 μL infiltration buffer, then add 25 mL of infiltration buffer to fill the ultracentrifuge tube and centrifuge again at 100,000 × *g* for 1 h. Resuspend the final pellet in 100 μL infiltration buffer and use EV samples immediately or store at −80°C (*see* **Note 14**).

3.5 Fungal EV Purification Using Discontinuous Sucrose Density Gradient Centrifugation

1. Prepare sucrose stocks from 10% to 90% (weight/volume), including 10%, 16%, 22%, 28%, 34%, 40%, 46%, 52%, 58%, 64%, 70%, and 90%, using sucrose and pure water.
2. Using a 15 mL ultracentrifuge tube on ice, layer 1 mL of each sucrose solution on top of one another, starting with the 90% sucrose solution on the bottom and sequentially up to the 16% sucrose solution (*see* **Note 15**).
3. Take 100 μL of fungal EVs isolated from liquid cultures or from *B. cinerea*-infected *Arabidopsis* leaves and mix it with 1 mL of the 10% sucrose solution, then load the mixture on top of the sucrose gradient.
4. Ultracentrifuge samples in a swing-bucket rotor for 16 h at 100,000 × *g* at 4°C.
5. Collect six fractions of 2 mL each from top to bottom, keeping fractions on ice, and weigh each fraction to calculate the density of each fraction (weight/2 mL).

6. For each fraction, transfer it to a new ultracentrifuge tube and dilute it with 12 mL of infiltration buffer and ultracentrifuge at 100,000 × *g* at 4°C for 1 h to obtain a final EV pellet. Remove the supernatant and resuspend the pellet in 20 μL of infiltration buffer.

4 Notes

1. New cultures should be obtained by inoculating plates directly from glycerol stocks rather than subculturing from existing plates because virulence can decrease through numerous rounds of subculturing.
2. For obtaining EVs, *B. cinerea* must be cultured very gently without rigorous shaking, because rigorous shaking will make the *B. cinerea* culture viscous, affecting following EV purification.
3. This method can be easily modified for use on other fungal species including *Verticillium dahliae*. To more easily process the culture supernatant, it may be necessary to add a step to filter the supernatant through sterilized double layers of Miracloth (Millipore) to remove mycelium.
4. For larger cultures or when using this protocol for other fungal species, the obtained supernatants can be concentrated approximately 20-fold using an Amicon ultrafiltration system (cutoff, 100 kDa). Then the concentrated supernatant can be centrifuged again at 10,000 × *g* at 4°C for 30 min to remove further debris.
5. Final EV pellets can be resuspended in any volume ranging from 10 to 100 μL depending on the starting amount of fungal tissue and desired concentration of EVs.
6. The maximum sample volume varies depending on the cartridge or EID size (small, medium, or large). The resuspension volume also varies by EID size (200, 400, and 1000 μL respectively).
7. This size cutoff allows particles smaller than 70 nm to diffuse through the membrane, while particles larger than 70 nm remain between the membranes for later resuspension.
8. Fill the water container with double-distilled water, and machine waste is already treated with bleach so it can be discarded down the drain.
9. EVs should only be frozen and thawed once before use; otherwise, lipid membrane integrity can be compromised.

10. It is critical not to let the infection progress more than 16h because after that point, plant cellular damage will begin to occur due to infection, and intracellular contaminants from plant cells will contaminate the fungal EV sample.
11. You will see the leaves become dark and slightly translucent when infiltrated.
12. Do not include any leaves that are ripped or damaged to avoid cellular contamination.
13. The roll should not be too tight; otherwise, the apoplastic wash fluid will not flow out, and it should not be too loose so that the tape does not fall down during centrifugation.
14. EV samples from the infected plant interface will contain a mixture of fungal and plant EVs; therefore, when analyzing the contents of these EVs, it is important to consider whether they are of fungal or plant origin.
15. Be gentle when loading sucrose layers on top of each other.

References

1. Liebana-Jordan M, Brotons B, Falcon-Perez JM, Gonzalez E (2021) Extracellular vesicles in the fungi kingdom. Int J Mol Sci 22(13):7221
2. de Toledo Martins S, Szwarc P, Goldenberg S, Alves LR (2019) Extracellular vesicles in fungi: composition and functions. Curr Top Microbiol Immunol 422:45–59
3. Rizzo J, Rodrigues ML, Janbon G. Extracellular vesicles in fungi: past, present, and future perspectives. Front Cell Infect Microbiol. 2020;10:346
4. Cai Q, Halilovic L, Shi T, Chen A, He B, Wu H, Jin H (2023) Extracellular vesicles: cross-organismal RNA trafficking in plants, microbes, and mammalian cells. Extracell Vesicles Circ Nucl Acids 4(2):262–282
5. Biquand E, Khau S, Papon N, Briard B (2024) Extracellular vesicles: new bullets in the fungal armory. Trends Microbiol 32(12):1167–1169
6. Rodrigues ML, Janbon G, O'Connell RJ, Chu TT, May RC, Jin H, Reis FCG, Alves LR, Puccia R, Fill TP, Rizzo J, Zamith-Miranda D, Miranda K, Gonçalves T, Ene IV, Kabani M, Anderson M, Gow NAR, Andes DR, Casadevall A,... Nimrichter L (2025) Characterizing extracellular vesicles of human fungal pathogens. Nat Microbiol 10(4):825–835
7. He B, Wang H, Liu G, Chen A, Calvo A, Cai Q, Jin H (2023) Fungal small RNAs ride in extracellular vesicles to enter plant cells through clathrin-mediated endocytosis. Nat Commun 14(1):4383
8. De Vallée A, Dupuy JW, Moriscot C, Gallet B, Vanderperre S, Guignard G, Rascle C, Calvar G, Malbert B, Gillet FX, Dieryckx C, Choquer M, Girard V, Poussereau N, Bruel C (2023) Extracellular vesicles of the plant pathogen *Botrytis cinerea*. J Fungi (Basel) 9(4):495
9. Escobar-Niño A, Harzen A, Stolze SC, Nakagami H, Fernández-Acero FJ (2023) The adaptation of *Botrytis cinerea* extracellular vesicles proteome to surrounding conditions: revealing new tools for its infection process. J Fungi (Basel) 9(9):872
10. Chen Y, Zhu Q, Cheng L, Wang Y, Li M, Yang Q, Hu L, Lou D, Li J, Dong X, Lee LP, Liu F (2021) Exosome detection via the ultra-fast-isolation system: EXODUS. Nat Methods 18(2):212–218

Chapter 15

Assays of Antifungal Activity of an Antimicrobial Peptide Against *Botrytis cinerea*

Min Zhang, Po Hu and Hailing Jin

Abstract

Antimicrobial peptides (AMPs) represent the first line of host immune defense against a broad spectrum of microorganisms. Whether naturally occurring or synthetically designed, these small peptides can potently inhibit microbial growth and remain stable under diverse conditions. Most AMPs act by directly disrupting fungal or bacterial membranes. In this study, we identified a glycine-rich peptide with strong antifungal activity against *Botrytis cinerea*. We further investigated its mode of action (MOA) and demonstrated its preventive antifungal activity in plants. Our work focuses primarily on the *in vitro* characterization of AMP features in the context of plant–*Botrytis* interactions.

Keywords Antimicrobial peptides (AMPs), Cell membranes, Glycine-rich, GUVs, Preventive antifungal activity, *Botrytis cinerea*, Fungal cell viability, Minimum inhibitory concentration (MIC)

1 Introduction

Gray mold, caused by *Botrytis cinerea* (*B. cinerea*), is a major postharvest disease that leads to the decay of fruits and vegetables, reducing both their quality and storage life. It also affects plant growth in the field [1,2], making the prevention and treatment of *B. cinerea* an urgent global challenge. For decades, gray mold management has relied primarily on synthetic fungicides [3]. However, excessive and prolonged use of these chemicals has raised serious concerns, including risks to human health, food safety issues, environmental pollution, and the emergence of resistant pathogen strains [3]. These challenges underscore the urgent need for novel, effective, and sustainable antifungal strategies to protect horticultural crops while ensuring both safety and productivity.

Antimicrobial peptides (AMPs) have emerged as promising alternatives to conventional fungicides. As the key components of innate defense systems in animals, plants, microorganisms, and insects [4], AMPs possess several advantageous properties, including short

Luca Nerva and Walter Chitarra (eds.), *Botrytis cinerea*, Methods in Molecular Biology, vol. 3076,
https://doi.org/10.1007/978-1-0716-5551-1_15,

length, high thermal stability, low toxicity, and a reduced possibility of resistance development [5]. In this study, we established a series of protocols ranging from AMP purification to antifungal activity assays against *B. cinerea*, providing a systematic framework to investigate plant–*B. cinerea* interactions. This approach holds significant potential for the development of AMP-based strategies to prevent gray mold infection in crops.

2 Materials

Prepare all solutions using distilled-deionized water (ddH_2O, Milli-Q) and analytical-grade reagents. Prepare and store reagents at the specified temperatures and adhere strictly to institutional safety and waste disposal regulations when handling or discarding materials.

2.1 Peptide Purification

1. LB broth (Luria–Bertani broth) medium: Add tryptone (5 g), yeast extract (2.5 g), and NaCl (5 g) into 500 mL of Milli-Q water, and autoclave (15 psi, 121°C, 15 min).
2. LB solid medium: 15% agar in the liquid medium and autoclave.
3. IPTG (1 M): Add 2.38 g of IPTG to 10 mL Milli-Q H_2O for a 1 M stock solution. Sterilize the IPTG stock solution through a 0.22 μm syringe filter and store as 1 mL aliquots at −20°C.
4. Antibiotic (Kanamycin): Make the stock concentration of 50 mg/mL, sterilize using a syringe filter, and store at −20°C as aliquots.
5. Binding/lysis buffer (NPI-10): 50 mM NaH_2PO_4, 300 mM NaCl, 10 mM imidazole, adjusting pH to 8.0 using NaOH and sterilizing using a 0.22 μm filter unit.
6. Washing buffer (NPI-20): 50 mM NaH_2PO_4, 300 mM NaCl, 20 mM imidazole, adjusting pH to 8.0 using NaOH and sterilizing using a 0.22 μm filter unit.
7. Elution buffer (NPI-250): 50 mM NaH_2PO_4, 300 mM NaCl, 250 mM imidazole. Adjust pH to 8.0 using NaOH and sterilize using a 0.22 μm filter unit.
8. 1× PBS buffer (0.01 M): 0.01 M Na_2HPO_4, 0.0018 M KH_2PO_4, 0.0027 M KCl, and 0.137 M NaCl, pH7.4, autoclave and store at room temperature.
9. Ni-NTA (nickel-nitrilotriacetic acid) agarose.
10. Poly-Prep chromatography column.
11. Zeba spin desalting column.
12. Amicon ultra centrifugal filter: 3 kDa MWCO.

2.2 Membrane Lipid Targets Screening

1. PVDF membrane or membrane lipid strips.
2. TBST (Tris Buffered Saline-Tween) buffer: Combine 100 mL of 10× TBS stock solution (200 mM Tris, 1.5 M NaCl, pH 7.6) with 900 mL of Milli-Q water, then add 1 mL of Tween-20 and mix well.
3. His-tag antibody.
4. Blocking buffer: 3% BSA in TBST.
5. Pierce ECL Western Blotting Substrate.

2.3 Giant Unilamellar Vesicles (GUVs) Assay

1. Ibidi μ-slide glass bottom 18-well chamber slide.
2. Corning cover glass: rectangular, No. 1, W × L 24 mm × 60 mm.
3. 1.15 mm spacer: single-sided sticky, rectangular well.
4. Medium butterfly binder clips: 1.25 inches.
5. Glass vials: 4 mL, screw top with polypropylene open-top cap.
6. Sucrose solution: 100 mM, ultrapure sucrose (≥99.5%), filtered with a 0.22 μm syringe filter. Store at 4°C.
7. Streptavidin solution: dissolve the lyophilized powder in Milli-Q water according to the manufacturer's instructions and store at 4°C.
8. 25 μL Hamilton Microliter Syringe with Cemented Needle.
9. Blue vinyl electrical tape.
10. 3 mL disposable syringe with 23 G needle.
11. Vacuum desiccator.
12. Slide rack.
13. Rehydration buffer (1× PBS buffer, 0.01 M): 0.01 M Na_2HPO_4, 0.0018 M KH_2PO_4, 0.0027 M KCl and 0.137 M NaCl, pH 7.4, autoclave and store at room temperature.
14. Prepare lipids in glass vials: dissolve lipids in chloroform to a final concentration of 10 mg/mL. Close the cap and seal the top with vinyl electrical tape to prevent moisture. Store at −20°C until use (*see* **Note 1**).
15. Prepare 5% (w/v) solution of polyvinyl alcohol (PVA, MW 145,000 g/mol): weigh 0.05 g PVA powder and put into a sterilized glass vial. Add 1 mL 100 mM sucrose solution, gently shake the mixture at RT for 30 min, then heat and stir at 85°C–90°C and 400 rpm for at least 8 h (or overnight if needed) until fully dissolved. Store at room temperature for up to 2 weeks (*see* **Notes 2 and 3**).
16. 3% BSA (w/v): dissolve BSA powder in Milli-Q water, sterilized by a 0.22 μm syringe filter. Store at 4°C.

2.4 Minimum Inhibitory Concentration (MIC) Assay

1. PDA (potato dextrose agar): add 24 g PDB to 1 L ddH_2O and stir. After thoroughly dissolving, add 15 g of agar and autoclave.
2. Minimal medium (2×): 60 g sucrose, 1 g $MgSO_4 \cdot 7H_2O$, 1 g KCl, 20 mg $FeSO_4 \cdot 7H_2O$, 4 g $NaNO_3$ in 1 L ddH_2O. Autoclave for 20 min and then add 0.4 mL trace elements stock solution.
3. The trace elements stock solution: 10 g citric acid, 10 g $ZnSO_4 \cdot 7H_2O$, 1 g $Fe(NH_4)_2(SO_4)_2 \cdot 6H_2O$, 0.5 g $CuSO_4 \cdot 5H_2O$, 0.1 g $MnSO_4 \cdot H_2O$, 0.1 g H_3BO_4 (boric acid), 0.1 g $NaMoO_4 \cdot 2H_2O$ in 100 mL ddH_2O. Sterile using 0.22 μm filter unit.
4. Resazurin solution: Prepare resazurin stock solution (10×) by dissolving 0.10 g resazurin sodium salt into 10 mL 1× PBS.
5. Corning 96-well polypropylene plate.

2.5 Fungal Cell Viability Assay

1. DMAO: 5 mM in DMSO.
2. EthD-III: 2 mM in Milli-Q H_2O.
3. 0.8% NaCl: 0.8 g NaCl in 100 mL ddH_2O.
4. Fungal inoculation medium: 0.1% Sabouraud maltose broth in ddH_2O. Use freshly made medium without autoclaving.

3 Methods

3.1 Peptide Purification

1. Revive transformed *E coli* BL21 cells (stored at –80°C) harboring the desired constructs in LB medium supplemented with proper antibiotic. Incubate at 37°C with overnight shaking.
2. Dilute the overnight culture at 1:100 into four 500 mL flasks containing LB medium with antibiotic. Incubate at 37°C with shaking until the OD600 reaches 0.6–0.8, then add IPTG to a final concentration of 100–500 μM to induce expression. For small peptides or proteins (<20 kDa), continue shaking at 37°C for 3–4 h to improve solubility and proper folding of the proteins.
3. Harvest the cells by centrifugation at 4000 × *g* for 20 min. Carefully discard the supernatant into a designated liquid waste container and treat it with bleach to ensure complete decontamination prior to disposal. Then freeze the cell pellets at −80°C until use (*see* **Note 4**).
4. Resuspend the cell pellets in 100 mL binding buffer. Disrupt the cells using an Emulsiflex C3 cell disruptor. High-pressure homogenization provides an efficient and reproducible method for cell disruption (*see* **Note 5**).

5. Clarify the cell lysate by centrifugation (10,000 × *g*, 15 min, 4°C) to remove cell debris and insoluble material. Collect the supernatant that contains soluble protein for downstream purification.
6. Binding: Pack 500 μL of Ni-NTA agarose (for His-tagged proteins/peptides) in a Poly-prep chromatography column and equilibrate with 5 mL of binding buffer by gravity flow. Apply 25 mL of the soluble protein sample to each column and allow it to pass through the resin by gravity. To increase the protein yield, the flow-through may be reapplied to the column.
7. Washing: Wash the column with 25 mL washing buffer by gravity flow to remove unbound proteins. Repeating the wash six to seven times improves the removal of nonspecifically bound proteins and enhances the purity of the target protein (*see* **Note 6**).
8. Elution: Elute the His-tagged protein using elution buffer containing a high concentration of imidazole. Apply 2 mL of elution buffer to each column and collect the flow-through containing the target protein.
9. Desalting: Exchange the buffer of the purified protein to 1× PBS using Zeba spin desalting columns, following the manufacturer's instructions.
10. Concentration: Concentrate the protein using an Amicon ultra centrifugal filter according to the manufacturer's instructions (*see* **Note 7**).
11. Detection and quantification: Analyze purified proteins by SDS-PAGE, followed by Coomassie blue staining. Determine protein concentration using the Bradford assay according to the manufacturer's instructions (*see* **Note 8**).

3.2 Membrane Lipid Targets Screening

1. Preparation of lipid-spotted membrane: Cut a PVDF membrane to 2 × 8 cm and mark the spot locations at least 1 cm apart for each lipid. Apply approximately 5 μg of each lipid slowly to the marked areas, forming 1–2 mm diameter spots. Allow the membrane to dry at room temperature for 30 min to 1 h. Or commercial lipid strips can be purchased from Echelon Biosciences Inc. and stored at 4°C.
2. Blocking: Incubate the membrane with 5–10 mL blocking buffer (3% BSA in TBST) and gently agitate for 1 h at room temperature or overnight at 4°C (*see* **Note 9**).
3. Protein incubation: Discard the blocking buffer and add the protein of interest (e.g., candidate peptide at 0.5 μg/mL) in 5 mL blocking buffer, ensuring the membrane is fully covered. Incubate for 1 h at room temperature with gentle agitation (*see* **Note 10**).

4. Washing: Discard the protein solution and wash the membrane three times with at least 5 mL TBST, 5–10 min each, with gentle agitation.
5. Antibody incubation: Prepare 5 mL blocking buffer containing 5 μL anti-His-HRP antibody. Incubate the membrane for 1 h at room temperature with gentle agitation (*see* **Note 11**).
6. Final washing: Wash the membrane three times with at least 5 mL TBST for 5–10 min each, with gentle agitation.
7. Detection: Detect target protein using an ECL detection system according to the manufacturer's instructions (*see* **Note 12**).

3.3 Giant Unilamellar Vesicles (GUVs) Assay

The following PVA-assisted GUV formation procedures (Fig. 1) are developed according to previous publications [6,7] with some modifications based on experimental practices. The critical hints for the successful results are highlighted in Sect. 4.

1. Coating the Coverslip: Spread 40 μL PVA solution evenly as a thin layer on a 24 × 60 mm coverslip using a 200 μL pipette tip (*see* **Notes 13 and 14**).
2. PVA layer drying: Dry the PVA layer at 50°C in an incubator for 30 min.
3. Lipid film deposition: Evenly spread 5 μL lipid solution on the PVA layer using a 25 μL glass syringe needle.

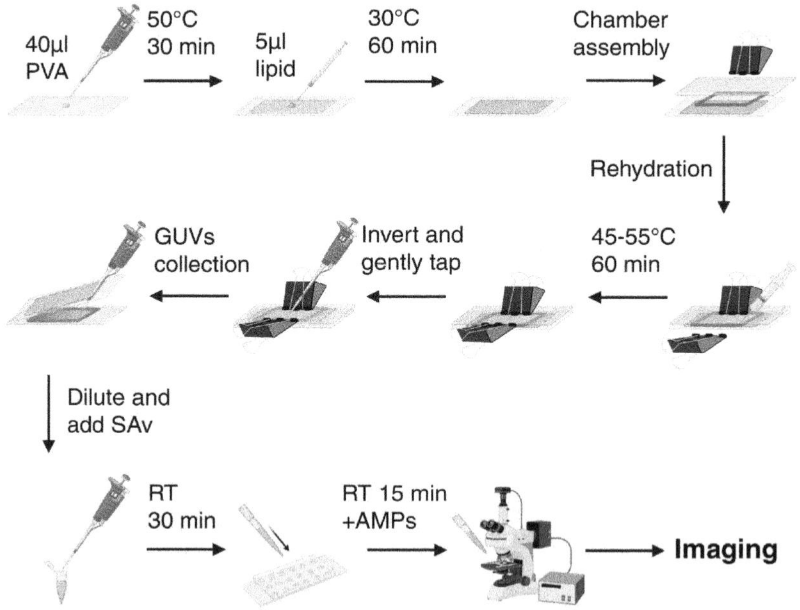

Fig. 1 Workflow of PVA-assisted giant unilamellar vesicles (GUVs) preparation. Created in https://BioRender.com

4. Lipid film drying: Dry the lipid-coated coverslip at 30°C in an incubator for 1 h. If using fluorescent-labeled lipid, protect the sample from light.
5. Chamber assembly: Assemble a chamber by placing a spacer (~30 × 22 × 2 mm) on the functionalized coverslip and cover it with a second coverslip. Secure the chamber with two butterfly clips on each side (*see* **Note 15**).
6. Rehydration: Add 1 mL of rehydration buffer (1× PBS) to the chamber using a 3 mL disposable needle syringe (23 G). Allow it to incubate at 45°C–55°C for 1 h to promote GUV formation (*see* **Note 16**).
7. Detachment: Gently invert the chamber and slightly tap on the glass surfaces (the coverslip with PVA layer and lipid film) with a pipette tip to detach the formed GUVs.
8. Harvesting: Keep the same orientation as in the last step, carefully remove the top coverslip (the coverslip with PVA layer and lipid film) to open the chamber. Collect the GUVs-containing solution with a 1 mL pipette and transfer it into a clean Eppendorf (EP) tube.
9. Settling: Allow the tube to stand at room temperature for half an hour (up to 2 h) to let the GUVs settle down before further use. Protect the samples from light if using fluorescent-labeled lipid. Place the tube on ice and use the GUVs on the same day (ideally) or keep them in the refrigerator for storage (up to 1 week).
10. Streptavidin treatment for adhesion: Add streptavidin (SAv) to freshly harvested GUVs (diluted for 2–10×) to a final concentration of 250 nM. Incubate for 30 min at room temperature.
11. Coating the glass-bottom chamber slide: Pretreat each well of an 18-well glass-bottom μ-slide chamber with 150 μL of 3% BSA for 10 min.
12. Washing the glass-bottom chamber slide: Remove the BSA solution and wash the wells with 150 μL Milli-Q water three times.
13. GUVs deposition: Remove the residual water and add 90 μL properly diluted and SAv-treated GUVs solution to each well. Wait for at least 15 min at room temperature to allow GUVs to settle down and attach to the glass bottom.
14. AMPs addition and time-lapse confocal imaging: Place the μ-slide chamber under the confocal microscope and find a single GUV or area of interest. Properly set the imaging parameters. Carefully add 10 μL of the AMPs (10×) from one corner of the well without disturbing the target GUV adhesion and start the time-lapse imaging (*see* **Note 17**).

3.4 Minimum Inhibitory Concentration (MIC) Assay

1. Fungal culture and spore preparation: *B. cinerea* strains are maintained on PDA plates at 22°C. Harvest the spores from 10-day-old cultures by flooding the plates with sterile water, then filter the suspension through a 70 μm cell strainer and centrifuge at 3000 × *g* for 2 min. Resuspend the resulting spores in 2× minimal medium and adjust to the desired density using a hemocytometer (*see* **Note 18**).
2. Peptide treatment in 96-well plates: Add 5000 spores in 50 μL 2× minimal medium to each well of a 96-well plate. Then add an equal volume of purified peptide to various final concentrations (0.25–8 μM) and incubate the plate for 48 h at 22°C–24°C with gentle shaking. PBS buffer mixed with fungal cells served as the "Quality Check" (QC) control (no peptide), and PBS buffer mixed with medium (no fungal cells) served as a negative control (*see* **Notes 19 and 20**).
3. Colorimetric readout: For the colorimetric assay, add 10 μL of resazurin to each well and incubate overnight at 37°C with gentle shaking. Wells with completely inhibited fungal growth will remain blue, similar to the negative control, whereas wells with fungal cell growth will turn pink or colorless, similar to the QC control (*see* **Note 21**).
4. Microscopic observation: Fungal growth inhibition can be further assessed and confirmed by microscopic observation.

3.5 Fungal Cell Viability Assay

1. Preparation of staining solution: Prepare a 100× staining solution by mixing 1 μL of DMAO and 2 μL of EthD-III with 8 μL of 0.8% NaCl. Vortex thoroughly to ensure complete mixing (*see* **Note 22**).
2. Fungal culture and spore collection: *B. cinerea* strains are maintained on PDA plates at 22°C. Spores can be collected from 10-day-old cultures by flooding plates with sterile water, filtering through a 70 μm cell strainer, and centrifuging at 3000 × *g* for 2 min.
3. Spore growth inhibition assay: Resuspend 5000 spores in 50 μL 1× PBS in a microcentrifuge tube. Add equal volume of the peptide at the desired concentration and incubate the mixture at room temperature for 1–3 h.
4. Hyphal growth inhibition assay: Resuspend 20,000 spores in 500 μL of 0.1% Sabouraud maltose broth medium and incubate overnight at room temperature for hyphal growth. After 16 h, centrifuge the cultures at 3000 × *g* for 2 min, discard the supernatant, and resuspend the hyphae in 50 μL 1× PBS. Add equal volume of the peptide at the desired concentration and mix thoroughly (*see* **Note 23**).
5. Staining and imaging: Add 1 μL of the 100× staining solution to 100 μL of cell culture and incubate for 30 min at room

temperature in the dark. Observe the stained cells using a confocal laser scanning microscope: DMAO can be detected with a FITC filter set, and EthD-III can be detected with a Texas Red or Cy3 filter set (*see* **Note 24**).

3.6 Inoculation-Based Antifungal Assay with B. cinerea

1. Fungal culture and spore preparation: *B. cinerea* strains are maintained on PDA plates at 22°C. Harvest spores from 10-day-old cultures by flooding the plates with sterile water, filtering through a 70 μm cell strainer, and centrifuging at 3000 × *g* for 2 min. Resuspend the spore pellets in 0.1% Sabouraud maltose broth and adjust the suspension to a final density of 1×10^5 spores/mL using a hemocytometer.
2. Plant sample preparation: Wash fruits and vegetables with ddH_2O and air-dry. Use rose petals directly without washing. Line the bottom of Ziploc boxes with double-layered paper towels and moisten with H_2O to maintain high humidity. Arrange the plant samples on the paper towels with sufficient spacing to avoid contact between samples.
3. Peptide treatment and inoculation: Evaluate antifungal activity using a drop-inoculation assay. Apply 10 μL of peptide solution or 1× PBS (negative control) onto the sample surfaces, followed immediately by an equal volume of spore suspension (final concentration 1×10^5 spores/mL). Alternatively, premix peptide with spores before applying to the sample surface.
4. Incubation and data collection: Incubate the samples under high-humidity conditions for 2–5 days, depending on the plant species. Monitor disease progression regularly and capture images for subsequent analysis.

4 Notes

1. Lipids can be used as pure components or as a mixture. Rhodamine-PE is 1.0 mol%, biotinyl-PE is 2.0 mol%. The lipid mixture can be stored in a well-sealed glass vial in the refrigerator for 2 weeks.
2. Fill a glass beaker with 1 L ddH_2O, microwave until boiling, then add a stir bar and place the beaker on a heated stir plate. Secure the PVA-containing glass vial to the wall of the beaker with tape. Monitor the temperature using a thermometer and maintain it at 85°C–90°C. Ensure that the PVA vial remains immersed in the water bath by checking every 1–2 h. During this period, briefly vortex the vial every 1–2 h to promote dissolution. A fully dissolved PVA solution should appear completely transparent with no visible clumps.

3. The molecular weight of PVA determines its melting and solidification properties. Avoid using lower-molecular weight PVA, as it will not solidify and dry properly. Residual moisture in the PVA gel layer can negatively affect lipid layer rehydration and GUV formation.
4. Bacterial pellets stored at −80°C tend to break apart more easily, which increases the protein yield.
5. Alternative cell disruption methods include ultrasonic treatment or the addition of lysozyme at an appropriate concentration. Avoid vigorous vortex after lysozyme addition to prevent protein denaturation.
6. Sufficient washing steps help remove unbound proteins and improve the purity of the target protein.
7. Highly Gly-rich peptides may precipitate and exhibit low solubility in PBS. To achieve higher concentrations, dry powder of the target protein may be obtained by methods such as co-precipitation with an amino acid carrier and/or lyophilization.
8. Use low-binding tubes (polypropylene) to minimize protein loss for the entire process.
9. During all incubation and wash steps, ensure that the membrane remains wet and does not dry out. Gentle agitation is recommended throughout these steps.
10. A starting concentration of 1–2 μg/mL is recommended for a new protein of interest. Avoid overnight incubation at 4°C as this may lead to protein degradation and reduced activity.
11. Use only purified proteins or antibodies; cell lysates are not recommended.
12. Avoid using ultrasensitive ECL reagents, as they may produce high background signals.
13. Coverslip cleaning: Prior to use, place coverslips on a rack and gently wash them in 70% ethanol for several hours. Dry and store them in a vacuum desiccator until use to keep PVA film dewetting.
14. An even and smooth PAV layer is critical for uniform lipid spreading and GUV formation. To achieve an optimal coating, place a 200 μL pipette tip horizontally at the center of the PVA drop and spread the solution evenly to both sides in a single motion, avoiding repeated back-and-forth movements. The thin PVA film remains sticky after spreading; excessive handling can result in uneven thickness.
15. Stack two 1 mm spacers with their adhesive sides facing each other to create a spacer with a final thickness of 2 mm.
16. Handle the chamber carefully and avoid shaking during rehydration. Optimal conditions may vary with lipid composition.

Good yields and vesicle sizes can be obtained under the following conditions: PC, PG, and PS—rehydrate with 1× PBS at 50°C; PE—rehydrate with 0.5× PBS at 50°C; PA—rehydrate with 0.5× PBS at 45°C.

17. Once loaded and attached to the glass-bottom chamber, GUV imaging should ideally be completed on the same day (within ~8 h). While some GUVs, like those composed of PC, PG and PS, remain relatively stable, others—particularly PE- and PA-GUVs—tend to deform by the following day.
18. Since we will add an equal volume of peptide to the cell culture, using 2× minimal medium is preferable.
19. The pH of the medium is critical for colorimetric assays; resazurin may appear pink under acidic conditions.
20. For Gly-rich peptides, use non-treated polypropylene 96-well plates to reduce peptide adhesion.
21. Fungal growth can be simply monitored by color changes or quantified using a spectrophotometer. After 3 h of incubation at 37°C, fluorescence can be measured using a plate reader (Ex: 530–560 nm, Em: 590 nm) [8,9].
22. Remove the growth medium prior to staining, as components such as nucleic acids in the medium can bind DMAO and EthD-III dyes, reducing staining efficiency.
23. For hyphal growth in 0.1% Sabouraud maltose broth, incubate cultures overnight without shaking. This slows growth and helps to promote optimal hyphal length.
24. DMAO exhibits higher staining efficiency in 0.8% NaCl. If possible, replace the PBS buffer with 0.8% NaCl prior to staining.

References

1. Case NT, Gurr SJ, Fisher MC et al (2025) Fungal impacts on Earth's ecosystems. Nature 638:49–57
2. Valero-Jiménez CA, Veloso J, Staats M, Van Kan JAL (2019) Comparative genomics of agrochemicals pathogenic *Botrytis* species with distinct host specificity. BMC Genomics 20:203. https://doi.org/10.1186/s12864-019-5580-x
3. Kim JO, Shin JH, Gumilang A et al (2016) Effectiveness of different classes of fungicides on *Botrytis cinerea* causing gray mold on fruit and vegetables. Plant Pathol J 32:570–574. https://doi.org/10.5423/PPJ.NT.05.2016.0114
4. Zhang L, Gallo RL (2016) Antimicrobial peptides. Curr Biol 26:R14–R19. https://doi.org/10.1016/j.cub.2015.11.017
5. Huan Y, Kong Q, Mou H, Yi H (2020) Antimicrobial peptides: classification, design, application and research progress in multiple fields. Front Microbiol 11:582779. https://doi.org/10.3389/fmicb.2020.582779
6. Iorio DD, Wegner SV (2024) Dynamic light-induced protein patterns at model membranes. J Vis Exp 204, e66531. https://doi.org/10.3791/66531
7. Weinberger A, Tsai FC, Koenderink GH et al (2013) Gel-assisted formation of giant unilamellar vesicles. Biophys J 105:154–164. https://doi.org/10.1016/j.bpj.2013.05.024
8. Tetorya M, Li H, Djami-Tchatchou AT et al (2023) Plant defensin MtDef4-derived antifungal peptide with multiple modes of action and potential as a bio-inspired fungicide. Mol Plant

Pathol 24:896–913. https://doi.org/10.1111/mpp.13336
9. Kharrat O, Yamaryo-Botté Y, Nasreddine R et al (2025) The antimicrobial activity of ETD151 defensin is dictated by the presence of glycosphingolipids in the targeted organisms. Proc Natl Acad Sci U S A 122(7):e2415524122. https://doi.org/10.1073/pnas.2415524122

Chapter 16

Evaluation of Biocontrol Efficacy by Competing Yeasts

Antonio Domenico Marsico, Margherita D'Amico and Falvia Angela Maria Maggiolini

Abstract

Botrytis cinerea causes gray mold across a wide range of crops and postharvest contexts. While integrated agronomic practices help, chemical fungicides are still often required, raising concerns about resistance development and environmental impact. Biological control agents (BCAs), particularly non-*Saccharomyces* yeasts, offer advantages such as simple nutritional needs, surface colonization potential, and lower risk of mycotoxin production. A robust, rapid, and cost-effective in vitro screening workflow is essential to identify promising BCAs that inhibit *B. cinerea* through fungistatic diffusible substances or volatile organic compounds (VOCs). This manuscript presents two complementary assays—the Cellophane Agar Layer technique (CALt) and the Sandwich Dual Culture technique (SDCt)—as practical tools for high-throughput screening and preliminary mechanism characterization. Methods are described in detail. CALt assesses the production of diffusible antifungal substances by yeasts by placing a cellophane layer over YPDA, applying yeast suspensions, and subsequently introducing *B. cinerea* to measure changes in fungal colony extension. SDCt evaluates the influence of yeast-emitted VOCs by sandwiching a *B. cinerea* mycelial disk between two plates containing 48-h-old yeasts and sealed to permit VOC diffusion. In both assays, daily mycelial growth of *B. cinerea* is recorded and analyzed as colony diameter growth over time, with results presented as mean ± standard deviation. The materials, media compositions (YPD and YPDA), and step-by-step protocols for yeast preparation, plate setup, inoculation, incubation conditions, and outcome calculations are provided to enable rapid replication and adoption in screening programs. Together, CALt and SDCt offer a streamlined framework to identify yeast BCAs with strong antagonistic activity and to gain preliminary insights into their mechanisms of action, informing subsequent validation under commercial conditions.

Keywords *Botrytis cinerea*, Biological control agents (BCAs), Non-*Saccharomyces* yeasts, High-throughput screening, In vitro screening

1 Introduction

Botrytis cinerea is an extremely polyphagous and ubiquitous pathogen and the causal agent of gray mold, both in field and postharvest conditions. Control of *B. cinerea* on various crops is commonly achieved with a combination of pesticide treatments and agronomic practices. Although very helpful, agronomic practices alone cannot

Luca Nerva and Walter Chitarra (eds.), *Botrytis cinerea*, Methods in Molecular Biology, vol. 3076,
https://doi.org/10.1007/978-1-0716-5551-1_16,

prevent the disease in many growing areas, so chemical treatments should normally be applied [1]. In recent years, the use of biological control agents (BCAs), such as bacteria, fungi, and yeasts, has been increased continuously, because of public concerns regarding the risk of pesticide residues in food and their negative impact on the environment [2]. An additional reason to reduce the use of synthetic chemical fungicides against *B. cinerea* is the rapid and relatively easy selection of resistant strains to single-site fungicides in fungal populations, caused by the continuous use of active substances with the same action mechanisms. Most of the BCAs described against *B. cinerea* are non-*Saccharomyces* yeasts, because they offer some advantages, such as simple nutritional requirements, the ability to colonize dry surfaces for long periods, and the inability to produce allergenic spores or mycotoxins [3]. The first step in developing BCAs is the screening process of strain collections, which will largely influence the efficacy and ultimately its success under commercial conditions. This chapter provides a guide to two rapid and cost-effective in vitro screening methods: Cellophane Agar Layer technique (CALt) and Sandwich Dual Culture technique (SDCt). Together, these tests not only allow us to select yeasts with strong antagonistic activity but also provide preliminary information on their mechanism of action. In particular, CALt evaluates the yeast's ability to produce fungistatic diffusible substances, while SDCt evaluates the ability to produce fungistatic volatile organic compounds (VOCs). The protocol of CALt was developed by Gibbs [4], while the protocol of SDCt was developed by Dennis and Webster [5]. Both have been used in several works to identify antagonistic yeast strains to *B. cinerea* [6,7].

2 Experimental Resouces

2.1 Materials

- Distilled water.
- Bacteriological peptone.
- Glucose.
- Yeast extract.
- Agar.
- Sterile Petri dishes.
- Glass loops.
- Cork borer, 9 mm in diameter.
- 50 mL sterile tubes.
- 1 mL sterile filter tips.
- Sterile cellophane disks.

2.2 Equipments

- 20°C dark incubator.
- Laminar airflow cabinet.

2.3 Media

Liquid Yeast Extract Peptone Dextrose (YPD) medium is prepared using the following recipe:

Component	Quantity
Bacteriological peptone	20 g
Glucose	20 g
Yeast extract	10 g
Distilled water	1000 mL

Yeast Extract Peptone Dextrose Agar (YPDA) medium is prepared using the following recipe:

Component	Quantity
Bacteriological peptone	20 g
Glucose	20 g
Yeast extract	10 g
Agar	15 g
Distilled water	1000 mL

3 Method

3.1 Cellophane Agar Layer Technique

- **Yeast preparation**
  - Grow yeast strains in flasks containing liquid YPD medium for 48 h.
  - Store at 4°C until use in the in vitro assays.
- **CALt layer preparation**
  - Place a 90 mm diameter sterile cellophane disk[1] onto solidified YPDA in culture plates[2].
  - Incubate the plates overnight to allow excess moisture to evaporate.
- **Yeast application**
  - Uniformly apply 50 μL of suspension containing 1.5×10^7 CFU mL^{-1} of each yeast onto the cellophane surface, using a sterile glass loop[3].

- Apply 50 μL of sterile water on a separate plate as a control.
- Prepare at least five replicates for each yeast strain and control.

- **Inoculation with *B. cinerea* and incubation**
  - After 2 days of incubation at 25°C in the dark, remove the cellophane disks (with or without yeast) from the plates[4] (Fig. 1).
  - Place a 9 mm diameter agar block, collected from the actively growing margin of *B. cinerea* using a cork borer, at the center of each plate.
  - Incubate again at 25°C in the dark.

Fig. 1 General workflow of Cellophane Agar Layer technique

- **Outcome measures**
  - Measure the growth extension of *B. cinerea* as colony diameter (mm) every 48 h of incubation, until the fungal colony in the control reaches the outer edge of the plate (Fig. 1).
  - Calculate the daily mycelial growth using the following formula:

$$\frac{\sum_{i=0}^{n}(y_{n+1} - y_n)}{N}$$

where:

y_n = diameter of the mycelium of *B. cinerea* at time n
y_{n+1} = diameter of the mycelium of *B. cinerea* at time $n + 1$
N = total number of days considered in the experiment

  - Output: mean ± standard deviation (SD) of *B. cinerea* daily mycelial growth for each yeast strain and control.

- **Notes:**
  [1] Cellophane disks can also be sterilized in an autoclave at 121°C for 15 min, taking care to place each disk between two sheets of absorbent paper and enclosing everything inside aluminum foil.
  [2] Make sure that the cellophane disk is well adhered to the entire surface of the agar substrate, avoiding any uncovered areas.
  [3] Accurate counting of spores is crucial for reproducibility, so it is necessary to vortex the spore suspension well before counting.
  [4] At this stage, it is important to check the substrate for yeast colonies. Discard any plates showing obvious yeast colonies.

3.2 Sandwich Dual Culture Technique

- **Yeast preparation**
  - Inoculate 50 μL of a suspension containing 1.5×10^7 CFU mL^{-1} of each yeast strain on Petri dishes containing YPDA.
  - Apply 50 μL of sterile water onto a separate plate as a control.
  - Prepare at least five replicates for each yeast strain and control.
  - Incubate at 25°C in the dark for 48 h.

- **Sandwich preparation**
  - Place a mycelial disk (9 mm diameter agar plug) of *B. cinerea* at the center of separate Petri dishes containing YPDA.
  - Individually coat the Petri dishes containing *B. cinerea* disk face-to-face with the plates containing 48-h-old yeast isolates to create a microenvironment suitable for the diffusion of volatile compounds possibly produced by the tested yeast isolates.

- Firmly wrap the two plates with Parafilm to prevent air leakage and incubate at 25°C in the dark[1].

- **Outcome measures**
  - Measure the growth extension of *B. cinerea* as colony diameter (mm) every 48 h of incubation, until the fungal colony in the control reaches the outer edge of the plate (Fig. 2).
  - Calculate the daily mycelial growth using the following formula:

$$\frac{\sum_{i=0}^{n} (y_{n+1} - y_n)}{N}$$

where:
y_n = diameter of the mycelium of *B. cinerea* at time n
y_{n+1} = diameter of the mycelium of *B. cinerea* at time $n + 1$
N = total number of days considered in the experiment

  - Output: mean ± standard deviation (SD) of *B. cinerea* daily mycelial growth for each yeast strain and control.

- **Notes:**
  [1] Place the *sandwiches* in the incubator with the plate containing the yeasts facing upwards.

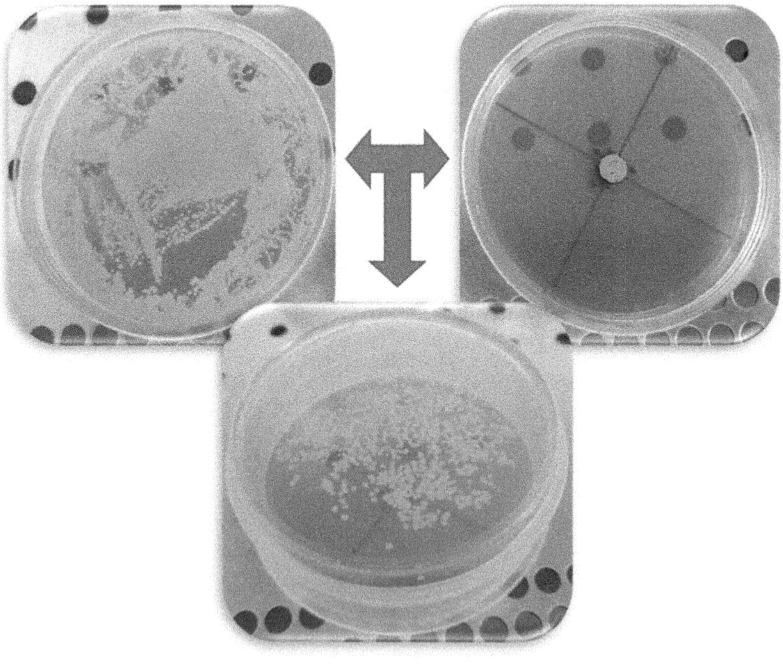

Fig. 2 General workflow of Sandwich Dual Culture technique

References

1. Jacometti MA, Wratten SD, Walter M (2010) Alternatives to synthetic fungicides for *Botrytis cinerea* management in vineyards. Aust J Grape Wine Res 16:154–162. https://doi.org/10.1111/j.1755-0238.2009.0067.x
2. Kowalska J, Krzymińska J, Tyburski J (2022) Yeasts as a potential biological agent in plant disease protection and yield improvement—a short review. Agriculture 12:1404. https://doi.org/10.3390/agriculture12091404
3. Sonker N, Pandey AK, Singh P (2016) Strategies to control post-harvest diseases of table grape: a review. J Wine Res. 27:105–122. https://doi.org/10.1080/09571264.2016.1151407
4. Gibbs JN (1967) A study of the epiphytic growth habit of *Fomes annosus*. Ann Bot 31:755–774. https://doi.org/10.1093/oxfordjournals.aob.a084180
5. Dennis C, Webster J (1971) Antagonistic properties of species-groups of *Trichoderma*: II. Production of volatile antibiotics. Trans Br Mycol Soc 57(1):41–48. https://doi.org/10.1016/S0007-1536(71)80078-5
6. Marsico AD, Velenosi M, Perniola R, Bergamini C, Sinonin S, David-Vaizant V, Maggiolini FAM, Hervé A, Cardone MF, Ventura M (2021) Native vineyard non-*Saccharomyces* yeasts used for biological control of *Botrytis cinerea* in stored table grape. Microorganisms 9:457. https://doi.org/10.3390/microorganisms9020457
7. Salerno A, D'Amico M, Bergamini C, Maggiolini FAM, Vendemia M, Prencipe A, Catacchio CR, Ventura M, Cardone MF, Marsico AD (2024) On the way to the technological development of newly selected non-*Saccharomyces* yeasts selected as innovative biocontrol agents in table grapes. Microorganisms 12:340. https://doi.org/10.3390/microorganisms12020340

Check for updates

Chapter 17

Methods for Evaluating the Efficacy of Microbial Biocontrol Agents Against *Botrytis cinerea* on Tomato Plant

Jean-François Bourgeay and Marc Bardin

Abstract

Botrytis cinerea, the causal agent of gray mold, is a highly adaptable necrotrophic fungus that causes significant losses in economically important crops such as tomatoes. Increasing resistance to fungicides and environmental concerns have stimulated interest in microbial biocontrol agents as part of integrated pest management strategies. This chapter presents standardized protocols for evaluating the efficacy of microbial biocontrol agents against *B. cinerea* on tomato plants under controlled conditions. Methods include preparation of pathogen and biocontrol inoculum, production of healthy plants, and bioassays on whole plants, detached stems, and detached leaves, with quantitative disease assessment. These protocols aim to ensure reproducibility and comparability while facilitating mechanistic studies and optimization of biocontrol strategies prior to field application.

Keywords *Botrytis cinerea* , Inoculation, Biocontrol, Efficacy, Plant, Stem, Leaves

1 Introduction

Botrytis cinerea is the causal agent of gray mold, a necrotrophic fungus that affects a wide range of crops, including crops of major economic importance such as tomatoes [1]. This pathogen is capable of producing large numbers of conidia that are easily dispersed by wind, can infect plants through various modes of attack, and can adapt to a wide range of control methods and environmental conditions, making it particularly difficult to control. Conventional management of gray mold has historically relied on repeated applications of fungicides. Their frequent use has contributed to the emergence of resistant strains [2] and raised environmental and regulatory concerns.

Biocontrol agents (BCAs), including beneficial microorganisms such as bacteria, yeasts, and filamentous fungi, have emerged as promising tools of integrated pest management (IPM) for *B. cinerea* [3]. They act through diverse mechanisms, including direct antagonism, such as antibiosis, mycoparasitism, and competition for nutrients and

Luca Nerva and Walter Chitarra (eds.), *Botrytis cinerea*, Methods in Molecular Biology, vol. 3076,
https://doi.org/10.1007/978-1-0716-5551-1_17,

space, as well as indirect, plant-mediated effects, including induced systemic resistance. They can also modify the local environment through the synthesis of volatile compounds, biosurfactants, organic acids, or redox changes, thereby inhibiting or slowing fungal growth and sporulation [4].

Experiments conducted under controlled conditions are necessary to demonstrate the potential of biocontrol agents to inhibit, delay, or reduce the development of *B. cinerea* in plants. These experiments enable rigorous evaluation of the intrinsic efficacy of the candidate biocontrol agent, elucidation of modes of action, characterization of dose–response relationships, optimization of application strategies prior to field deployment based on various factors (e.g., microclimatic conditions, plant fertilization levels, etc.), and effective integration into integrated crop protection programs (e.g., compatibility with other treatments, side effects, etc.).

This chapter provides protocols for evaluating the efficacy of microbial biocontrol agents against *B. cinerea* on tomato plants under controlled conditions. These protocols include the following steps:

(1) culture and inoculum preparation of *B. cinerea*,

(2) preparation of biocontrol agents,

(3) plant production and inoculation with the pathogen and treatment with biocontrol agents on whole plants and detached leaves,

(4) symptom assessment.

The protocols presented aim to standardize the conditions for inoculation, incubation, and symptom evaluation in order to ensure the comparability and reproducibility of results.

2 Culture and Inoculum Preparation of *B. cinerea*

Botrytis cinerea isolates used for plant testing can come from different host plants or be isolated outside of an agricultural context [5]. Prior to experimental use, all strains were purified by single spore isolation to ensure their genetic homogeneity. In most studies evaluating the efficacy of biocontrol agents, we use the monoconidial strain of *B. cinerea*, BC1, isolated from a glasshouse-grown tomato in Plougastel, France, in 1989. This strain was selected for its high level of aggressiveness on tomato plants in preliminary studies (following the whole-plant protocol identical to that described in this chapter) [5, 6]. After purification, *B. cinerea* strains are stored in cryotubes at −80°C either as spore suspensions in a 20% (vol/vol) phosphate–glycerol buffer or as mycelial plugs without added buffer. To produce inoculum, cryotubes containing spore suspensions or mycelial plugs are defrosted at room temperature and then homogenized by vortexing.

The strains are then grown on a pre-prepared potato dextrose agar (PDA) medium, sterilized in an autoclave, and distributed in Petri dishes. The Petri dishes containing PDA were inoculated using two complementary methods.

For spore production (Fig. 1), three drops of spore suspension (≈2 μL) are placed on the surface of the agar medium in a 90-mm Petri plate, forming an equidistant triangle to ensure uniform colony growth. The conidia are then harvested from the surface of the culture medium after 10–14 days of incubation under cool white fluorescent light (14-h photoperiod, 114 μmol m^{-2} s^{-1}) in a growth chamber at 21°C. The spores are harvested by scraping the surface of the PDA medium with a sterile Drigalski spatula in the presence of 4 mL sterile osmosis water, then filtered through a 30 μm mesh filter (Nylon Net Filter) to remove mycelial fragments. The spore suspension is vortexed for 1 min with 5 g of glass beads (2.5 mm diameter) to separate the spores. The conidia concentration in the suspension is then determined using a hemocytometer (Malassez cell) under a microscope (10× magnification) and adjusted to the desired concentration of conidia. To verify the viability of the conidia used during an experiment, 100 μL of the conidial suspension is spread onto PDA medium and incubated for 24 h at laboratory room temperature (≈20°C). Conidial germination is then assessed by microscopic observation (10× magnification) by counting 100 conidia per Petri dish. The germination rate is then expressed as the percentage of germinated conidia.

For mycelial production, a drop of spore suspension (≈2 μL) is placed in the center of a Petri dish containing PDA medium to promote uniform mycelial growth (Fig. 2). Strains are grown for 3 days in a growth chamber (21°C, 14 h of photoperiod at 114 μmol m^{-2} s^{-1}). Mycelial agar plugs of 5 mm diameter taken from the growing margin of the 3-day-old culture were used as inoculum.

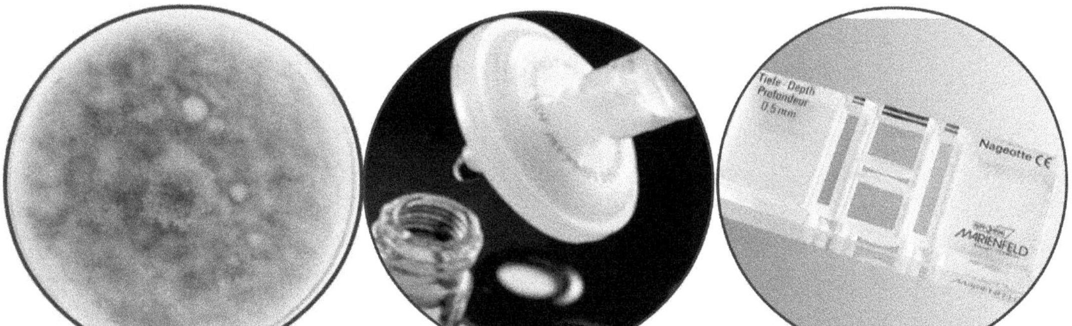

Fig. 1 Culture of *Botrytis cinerea* showing dense gray sporulation after 14 days of incubation at 21°C on PDA medium, filtration of the spore suspension, and a hemocytometer (Malassez cell) to adjust the spore concentration

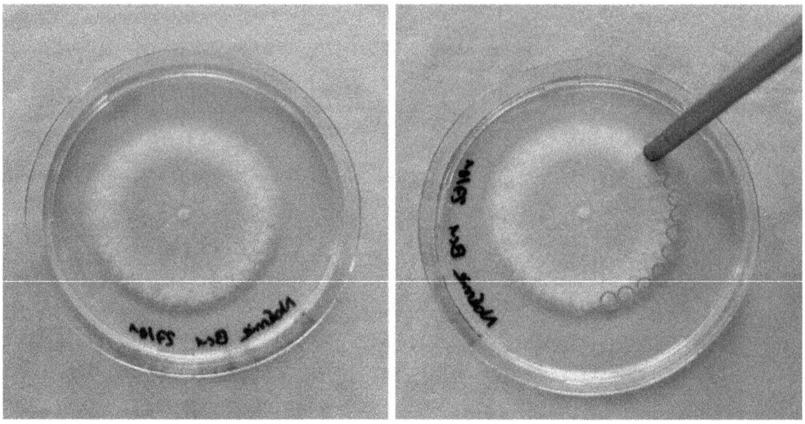

Fig. 2 Culture of *Botrytis cinerea* showing dense mycelial growth after 3 days of incubation at 21°C on PDA medium. Preparation of mycelial plugs from the margin of the culture medium

The inoculated Petri dishes are incubated in a growth chamber at a constant temperature of 21°C with a 12-h photoperiod. All manipulations (preparation of media, plating, and inoculations) are carried out under a horizontal laminar flow hood, under strict aseptic conditions, including disinfection of work surfaces and hands, as well as the use of sterile equipment. After use, reusable materials are disinfected with either alcohol or bleach.

3 Preparation of Microbial Biocontrol Agents

For the preparation of microbial biocontrol agents, there are two possible scenarios:

- The microbial biocontrol agent is formulated and potentially commercially available in the form of a product. In this case, preparation consists of following the recommendations of the manufacturer according to the properties of the formulated product (e.g., liquid suspension, wettable powder, etc.). If one of the objectives of the experiment is to test different doses of the product, the dose approved or recommended by the manufacturer is used as a reference, serving as a standard for comparison.
- The microbial biocontrol agent is a strain of microorganism that is not formulated in a product. In this case, the principles of basic microbiology apply to its preparation (i.e., culture on agar medium, incubation, collection of living cells, adjustment to the desired dose, etc.). In this context, the choice of the dose to be applied is often a key step in the experimental protocol, as it

determines the optimal dose required to achieve maximum protection from the microbial biocontrol agent. Reduced doses can be tested to identify the minimum effective dose of the agent, while contributing to cost reduction and limiting environmental impact. Conversely, higher doses can be incorporated to identify the threshold for the occurrence of unfavorable effects and to detect the possible development of phytotoxicity. In this case, concentrations are expressed in biological units (generally $CFU{\cdot}mL^{-1}$ for bacteria or $spores{\cdot}mL^{-1}$ for fungi). The viability of the biocontrol inoculum used during the experiments is verified. In the case of a biocontrol fungus, the same procedure as that used for *B. cinerea* is applied (measurement of spore germination on agar medium, see above). In the case of bacteria, the dilution–plating technique is used (determination of the number of CFUs). All operations are carried out under strict aseptic conditions including use of a horizontal laminar flow hood, disinfection of work surfaces and hands, and use of sterile equipment. After use, reusable materials are disinfected with either alcohol or bleach.

Microbial biocontrol agents or products are applied either by foliar spraying, by direct application onto the target, or at the root level in the soil or in the culture substrate of plants (e.g., rockwool, compost, etc.), depending on their modes of action and the objectives of the experiment. Treatments can be applied before inoculation with *B. cinerea* (preventive effect), simultaneously with inoculation, or after inoculation (curative effect), depending on the objectives of the study.

4 Plant Production

Tomato plants (*Solanum lycopersicum*) of a variety known to be susceptible to *B. cinerea* are used. Various tomato varieties susceptible to gray mold are available on the market. In our experiments, we generally use the susceptible varieties Clodano or Climberley (Syngenta Vegetable Seeds), and Rose de Berne or Monalbo (INRAE). The plants are grown in a greenhouse in individual pots (10 × 10 cm) containing commercial peat compost (Klasmann-Deilmann, ref. 416 TS 3 fine, special for sowing) and watered daily with a commercial nutrient solution (TIMASOL, NPK 15–10–30). The level of nitrogen fertilization is controlled during plant growth, as this can influence the susceptibility of plants to *B. cinerea* [7]. To conduct the experiments, plants aged 6–10 weeks with six to ten leaves are used. The age of the plants to be used depends on the microclimatic conditions prevailing in the greenhouse, the desired plant size, and the number of leaves required. For example, in a greenhouse located in a temperate Mediterranean climate (southern France), plants are generally grown for 10 weeks in winter and 6 weeks in summer.

5 Plant Bioassays

Three types of bioassays are performed to assess the efficacy of microbial biocontrol agents to protect tomato against *B. cinerea*.

5.1 Whole Plant Test

To evaluate the effect of microbial biocontrol agents on *B. cinerea* stem canker development, a whole plant test can be conducted [8]. At the time of inoculation with the pathogen, plants aged 6–10 weeks are transferred to a growth chamber under the following controlled conditions: 22°C during the day, 20°C at night, a photoperiod of 14 h with a light intensity of 162 μmol m^{-2} s^{-1}, and relative humidity above 85% in order to place *Botrytis* in the most favorable conditions for infestation of the stem. Tomato leaves are cut with disinfected scissors from the plants, leaving between 5 and 10 mm of petiole stub on the stem. Leaving longer petiole stubs (> 10 mm) can reduce the incidence of the disease [9]. This decrease in the infection rate is attributed to the abscission and falling of infected petiole stubs before the pathogen can colonize them completely and reach the stem tissue. This abscission can also be observed with shorter stubs, but their reduced size often allows the pathogen sufficient time to colonize the stem.

On plants aged 8 weeks with seven to eight developed leaves, two to three leaves are excised (typically leaves numbered 3, 5, and 7 from oldest to youngest). A volume of 10 μL of conidia suspension is then deposited on each pruning wound using a micropipette (Fig. 3). For these tests, the optimal dose of *B. cinerea* conidia to be deposited on the wounds in order to obtain good reproducibility of results was estimated, in previous laboratory studies, at 10^4 conidia per wound (i.e., 10 μL of a suspension dosed at 10^6 conidia mL^{-1}).

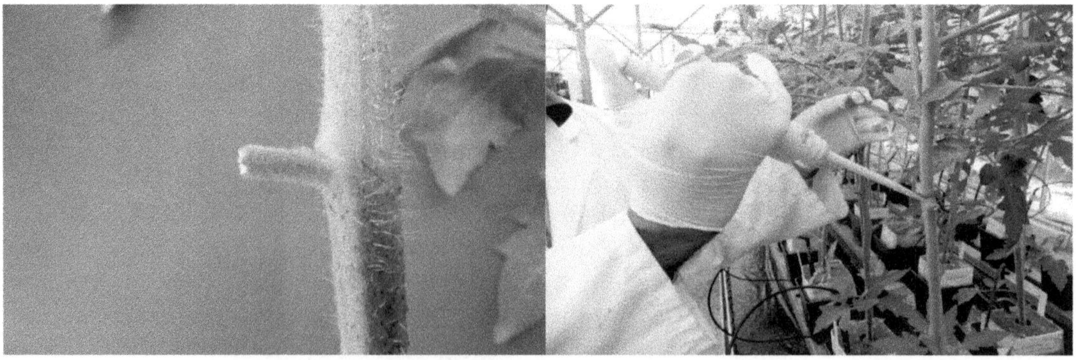

Fig. 3 Petiole stub remaining after cutting the tomato leaf and inoculating a plant with a conidial suspension of *B. cinerea* on the pruning wound using a micropipette

The viability of the conidia used in the experiments is verified by a spore germination test as described above.

The microbial biocontrol agent is then applied to pruning wounds with 10 μL of a cell suspension (e.g., bacterial cells, spores) dosed at the desired concentration, after complete absorption by the petiole of the inoculum of *B. cinerea* (approximately 5 min). As for the pathogen, the viability of the biocontrol inoculum used during the experiments is verified as described above. To test a potential preventive effect of the microbial biocontrol agents against *B. cinerea* on tomato plants, inoculation of *B. cinerea* is realized 0, 2, 4, or 6 h after treatment with the selected microbial biocontrol agent. To test a potential curative effect of the microbial biocontrol agents against *B. cinerea*, the same method is used except that inoculation of *B. cinerea* at 10^6 spores mL^{-1} is realized before the treatment with the microbial biocontrol agent. In this case, treatment with the microbial biocontrol agent is realized 0, 2, 4, or 6 h after the inoculation with *B. cinerea*. The control plants (untreated plants) are only inoculated with the conidial suspension of the pathogen. Treated and untreated plants are incubated in a growth chamber under the controlled conditions described above. Plants are randomly distributed throughout the climate-controlled chamber to minimize environmental biases.

The experiment is generally repeated three times independently, each with three to five replicates (three to five plants, each with two to three pruning wounds) per treatment. The size of the lesion in mm on the tomato stem is measured each day visually using a millimeter ruler, from the third to the seventh day after inoculation (Fig. 4).

Fig. 4 Example of *B. cinerea* stem lesions on a tomato plant after inoculation with a conidial suspension of *B. cinerea* on the pruning wound

To take into account the kinetics of disease development, we computed the area under the disease progress curve (AUDPC) as described by Decognet et al. [6]. To compare the efficacy of the BCAs, a protection index is computed as $100 \times (AUDPC_{BC} - AUDPC_{biocontrol})/AUDPC_{BC}$, where $AUDPC_{BC}$ corresponds to the average AUDPC values of plants inoculated with *B. cinerea* and left untreated, and $AUDPC_{biocontrol}$ corresponds to the average AUDPC values of plants inoculated with *B. cinerea* and treated with the biocontrol agent.

Other qualitative variables are measured in the absence of lesions on the stem: 0 = no symptoms on the petiole stub; 1 = necrosis on part of the petiole stub; 2 = petiole stub completely necrotic; * = petiole stub fallen from the stem.

5.2 Detached Tomato Stem Test

To evaluate the effect of microbial biocontrol agents on *B. cinerea* stem canker development and the sporulation capacity of *B. cinerea*, a bioassay can be conducted using detached tomato stems [10,11]. To this end, 40- to 80-mm stem segments are excised from 6- to 10-week-old tomato plants produced as described in the plant production part (Fig. 5). Each stem segment includes one leaf, which is cut, leaving a 5- to 15-mm petiole stub on the stem as described above. To assess pathogen sporulation, the petiole fragments should be weighed (fresh weight, g) prior to inoculation and treatment. Segments are then placed individually in 55-mm-diameter screw-cap tubes or into 24-well plates and all arranged in transparent polystyrene boxes containing moistened absorbent filter paper, covered with lids, to maintain high relative humidity (approximately 100%). The pruning wounds are inoculated with a conidial suspension of *B. cinerea* following the same procedure described above for whole tomato plants. The microbial biocontrol agent is then applied to pruning wounds with 10 μL of a cell suspension (e.g., bacterial cells, spores) dosed at the desired concentration, 5 min after the inoculation

Fig. 5 Biocontrol agent efficacy test on 40-mm detached tomato stems. On the left, the control showing lesions on tomato stems after inoculation of a suspension of *B. cinerea* conidia on the pruning wound, 4 days after inoculation. On the right, sections of stems protected by a biocontrol agent

of *B. cinerea*, the time that the drop is absorbed by the petiole. The viability of the pathogen inoculum and that of the biocontrol agent is verified as described above. The inoculated stem segments were incubated in a growth chamber under controlled conditions (21°C, photoperiod L:D 14:10, light intensity 114 μmol m^{-2} s^{-1}) in order to favor the development of *B. cinerea* on the stem.

For each concentration of a given microbial biocontrol agent, replicate stem samples are inoculated, and the experiment was generally repeated three times independently. Lesion length (mm) is measured each day visually using a millimeter ruler, from the third to the seventh day after inoculation. As for the whole plant test, the AUDPC from the third day to the seventh day after inoculation is calculated, and a protection index of tomato generated by the microbial biocontrol agent is computed.

Conidial production by *B. cinerea* can be quantified 10 days after inoculation, following the method of Abro et al. [12]. Stem samples are removed from the growth chamber, and 5 mL of an alcoholic solution (containing 20% ethanol [vol/vol] and 100 μL of Tween 80 per liter) is added to each cap. A centrifuge tube is then attached to each cap, and the samples are stored at 4°C until spore counting. Two liquid subsamples are taken from each tube, and spores are counted using a Malassez hemocytometer. Spore production for each detached organ is expressed as the number of spores per gram of fresh weight, and the percentage of sporulation inhibition relative to the control is calculated.

5.3 Detached Leaf Test

The detached leaf assay is a higher-throughput method used to rapidly assess the aggressiveness of *B. cinerea* and the efficacy of microbial biocontrol agents (Fig. 6). The protocol involves inoculation with mycelial plugs of the pathogen and quantification of lesion expansion using image analysis [13]. For this assay, leaves are sprayed with a suspension of the microbial biocontrol agent (e.g., bacterial cells or fungal spores) at the desired concentration. After drying for 30–60 min, leaves are excised the day of inoculation, and two leaflets are placed in transparent plastic boxes lined with moistened absorbent

Fig. 6 Biocontrol agent efficacy test on detached tomato leaves. Inoculation of leaflets by placing the mycelial plugs in the center of the leaflets, and lesions caused by *B. cinerea* on the leaflets

paper to maintain high relative humidity (approximately 100%). A preventive effect of the treatment can be evaluated. In this case, treatment with the selected biocontrol agent is carried out before inoculation by the pathogen (e.g., several days later).

Mycelial plugs (5 mm in diameter) of *B. cinerea*, excised from the actively growing margin of 3-day-old PDA cultures, are placed in the center of the tomato leaflets, with the mycelium in direct contact with the plant tissue. For each treatment, six to ten leaflets (distributed across three to five transparent plastic boxes) are inoculated. The boxes are closed, and the detached leaves are incubated in a growth chamber under conditions favorable for disease development (21 °C, 14 h photoperiod at 114 μmol s^{-1} m^{-2}). Leaflets from untreated plants serve as controls. The experiment is generally performed in three independent replicates.

Detached leaves are photographed every day, from day 3 to day 6 after inoculation, using a digital camera. The *B. cinerea* lesion areas are assessed with ImageJ software (US National Institutes of Health, Bethesda, MD, USA), and the necrotic areas measured are compared between treatments. The AUDPC can be calculated as described above. To compare the protection induced on the leaves by the different treatments, a protection index is computed as described above.

6 Conclusion

The protocols described in this chapter provide a robust and reproducible methodological framework for studying the aggressiveness of *B. cinerea* and for evaluating the efficacy of microbial biocontrol agents, both on whole plants and on detached organs (stems or leaves), under controlled conditions. These approaches enable detailed quantitative analysis of host–pathogen–biosolution interactions. The same types of experimental approaches can be applied to test the level of aggressiveness of different isolates of *B. cinerea* [5, 14, 15] and/or other biosolutions, such as natural plant extracts, microbial extracts, or pure metabolites [16]. These approaches can also be used on other plant species and with other pathogens [17, 18].

Acknowledgement

These protocols were improved, refined, and evaluated as part of the priority research program CapZeroPhyto (Investissements d'avenir, Cultiver et protéger autrement ANR-20-PCPA-0003), and through the ADOPT-IPM project, funded by the European Union's Horizon Europe Framework Program (Grant Agreement No. 101060430).

References

1. Bardin M, Gullino ML (2020) Fungal diseases. In: Gullino ML, Albajes R, Nicot PC (eds) Integrated pest and disease management in greenhouse crops. Plant pathology in the 21st century. Springer, pp 55–100.
2. Sofianos G, Samaras A, Karaoglanidis G (2023) Multiple and multidrug resistance in *Botrytis cinerea*: molecular mechanisms of MLR/MDR strains in Greece and effects of co-existence of different resistance mechanisms on fungicide sensitivity. Front Plant Sci 14:1273193. https://doi.org/10.3389/fpls.2023.1273193
3. Nicot PC, Stewart A, Bardin M, Elad Y (2015) Biological control and biopesticide suppression of Botrytis-incited diseases. In: Fillinger S, Elad Y (eds) Botrytis – the fungus, the pathogen and its management in agricultural systems. Springer, pp 165–187
4. Turc E, Pressecq T, Nicot PC, Bardin M (2023) Modes of action of microbial biocontrol agents against plant diseases. In: Puopolo G (ed) Microbial biocontrol agents: developing effective biopesticides. CAB International, pp 45–68. https://doi.org/10.1079/9781789249200.0003
5. Bardin M, Leyronas C, Troulet C, Morris CE (2018) Striking similarities between *Botrytis cinerea* from non-agricultural and from agricultural habitats. Front Plant Sci 9:1820. https://doi.org/10.3389/fpls.2018.01820
6. Decognet V, Bardin M, Trottin-Caudal T, Nicot PC (2009) Rapid change in the genetic diversity of *Botrytis cinerea* populations after the introduction of strains in a tomato glasshouse. Phytopathology 99:185–193
7. Abro MA, Lecompte F, Bardin M, Nicot PC (2014) Nitrogen fertilization impacts biocontrol of tomato gray mold. Agron Sustain Dev 34:641–648. https://doi.org/10.1007/s13593-013-0168-3
8. Bardin M, Fargues J, Nicot PC (2008) Compatibility between biopesticides used to control grey mould, powdery mildew and whitefly on tomato. Biol Control 46:476–483. https://doi.org/10.1016/j.biocontrol.2008.05.012
9. Decognet V, Ravetti F, Martin C, Nicot PC (2010) Improved leaf pruning reduces development of stem cankers caused by grey mould in greenhouse tomatoes. Agron Sustain Dev 30:465–472. https://doi.org/10.1051/agro/2009030
10. Foughalia A, Hamiroune M, Chandeysson C, Bourgeay JF, Duffaud M, Bardin M, Nicot PC, Aissat K (2024) Biocontrol potential of native Algerian bacteria isolated from greenhouses against *Botrytis cinerea* and *Oidium neolycopersici* on tomato. J Plant Dis Prot 131:847–861. https://doi.org/10.1007/s41348-024-00865-6
11. Bouaoud Y, Troulet C, Foughalia A, Berge O, Aissat K, Bardin M (2018) A multi-criteria approach for the selection of efficient biocontrol agents against *Botrytis cinerea* on tomato in Algeria. BioControl 63:299–311
12. Abro MA, Lecompte F, Bryone F, Nicot PC (2013) Nitrogen fertilization of the host plant influences production and pathogenicity of *Botrytis cinerea* secondary inoculum. Phytopathology 103:261–267
13. Boukaew S, Prasertsan P, Troulet C, Bardin M (2017) Biological control of tomato gray mold caused by *Botrytis cinerea* by using *Streptomyces* spp. BioControl 62:793–803. https://doi.org/10.1007/s10526-017-9825-9
14. Mercier A, Simon A, Lapalu N, Giraud T, Bardin M, Walker AS, Viaud M, Gladieux P (2021) Population genomics reveals molecular determinants of specialization to tomato in the polyphagous fungal pathogen *Botrytis cinerea* in France. Phytopathology 111:2355–2366. https://doi.org/10.1094/PHYTO-07-20-0302-FI
15. Monteil C, Bardin M, Morris CE (2014) Features of air masses associated with the deposition of *Pseudomonas syringae* and *Botrytis cinerea* by rain and snowfall. ISME J 8:2290–2304. https://doi.org/10.1038/ismej.2014.55
16. Zapletalová M, Rancurel C, Industri B, Bardin M, Le Brigand K, Nicot PC, Magnone V, Seassau A, Barbry P, Potěšil D, Zdráhal Z, Ponchet M, Lochman J (2023) BABA-induced pathogen resistance: a multi-omics analysis of the tomato response reveals a hyper-receptive status strongly involving ethylene. Hortic Res 10:1–14. https://doi.org/10.1093/hr/uhad068
17. Forges M, Bardin M, Urban L, Aarrouf J, Charles F (2020) Impact of UV-C radiation applied during plant growth on pre- and post-harvest disease sensitivity and fruit quality of strawberry. Plant Dis 104:3239–3247. https://doi.org/10.1094/PDIS-02-20-0306-RE
18. Leyronas C, Bardin M, Berthier K, Duffaud M, Troulet C, Torres M, Villeneuve F, Nicot PC (2018) Assessing the phenotypic and genotypic diversity of *Sclerotinia sclerotiorum* in France. Eur J Plant Pathol 152:933–944. https://doi.org/10.1007/s10658-018-1493-9

Chapter 18

Procedures for the Study of *Botrytis cinerea* Proteome

Almudena Escobar-Niño, Olivier Coste, Maria Victoria Calcis-Marzán and Francisco Javier Fernandez-Acero

Abstract

Proteomics has been revealed as a key set of technologies that provide a detailed description of the molecular processes involved in the development of a specific phenotype. "Omics" technologies can collect an incredible amount of information. Among them, proteomics is an invaluable tool for defining specific biological information by studying the complete set of proteins under specific conditions, the proteome; or specific subsets of proteins, the subproteome. It is a crucial instrument for describing protein post-translational modifications, the functional annotation of the genome, and the detection of orphan genes. Protein extraction procedures are necessary to obtain *B. cinerea* protein extracts of sufficient quality to be analyzed by LC-MS/MS, avoiding contaminants that interfere with the identification process. After experimental design, collect the samples and replicates as defined in each experimental approach; we will describe protocols and procedures for the next steps of proteome and subproteome extraction and LC-MS analysis.

Keywords Proteome, Subproteome, LC-MS, Orbitrap

1 Introduction

Since 2006, when the first proteomics approach to *Botrytis cinerea* was published [1], and its potential and applications were soon highlighted [2, 3], our group has been involved in the development of protocols that allow the development of scientific approaches to the biology and pathology of this model phytopathogenic fungus, especially from the protein point of view.

From the initial goals based on 2DE gels, the efforts have focused on the preparation of new approaches based on liquid chromatography separation coupled with mass spectrometry instruments (Fig. 1). In spite of these improvements, the global coverage of the obtained proteome compared with the annotated genes in the sequencing projects is still poor. Our data show that, after almost 20 years of uncovering fungal proteins, we have barely progressed beyond 52% of the genome [4].

Luca Nerva and Walter Chitarra (eds.), *Botrytis cinerea*, Methods in Molecular Biology, vol. 3076,
https://doi.org/10.1007/978-1-0716-5551-1_18,

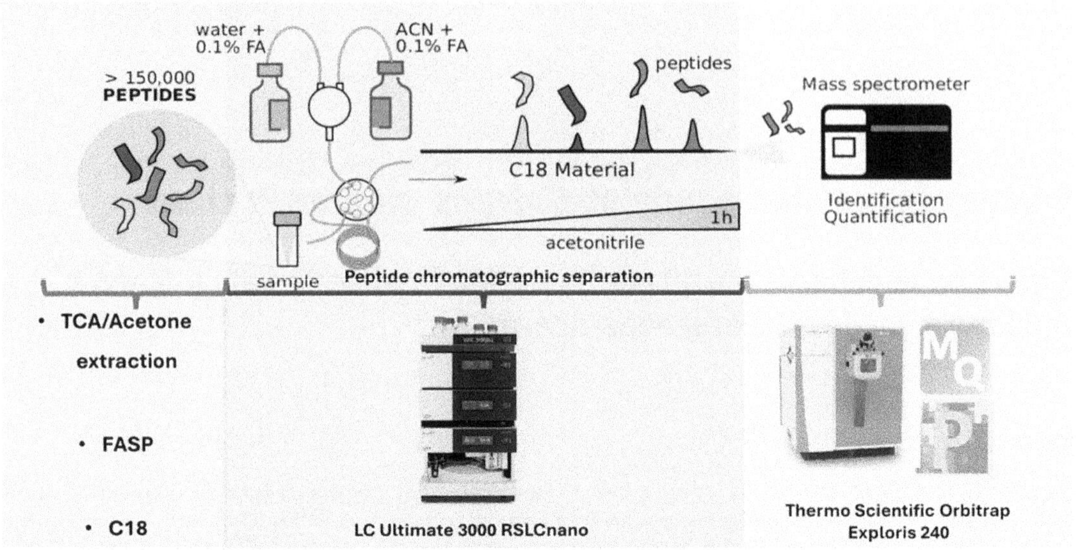

Fig. 1 General workflow of proteomics approaches, from protein/peptide extraction, digestion, separation, and analysis

In this chapter, we will share our latest procedures and protocols to produce proteomic approaches for *B. cinerea*, including cultures and infections, protein extraction and cleaning, trypsin digestion, and LC-MS/MS analysis preparation (**See Notes 1-5**).

2 Materials

2.1 Fungal Strains

The fungal strain *B. cinerea* B05.10 was provided by the Spanish Type Culture Collection (CECT).

2.2 Equipment

- 15 mL and 50 mL Falcon tubes (low-binding, preferably polyvinyl or polypropylene).
- 1.5 mL and 2 mL microcentrifuge tubes (low-binding, preferably polyvinyl or polypropylene).
- Low-binding pipette tips (preferably polyvinyl or polypropylene).
- COREX tubes.
- LC-MS vials and adapters.
- 30 μm nylon filters.
- 0.22 μm and 0.45 μm filters (PES or cellulose).
- Ultrafiltration membranes (PES or cellulose), 100 kDa cutoff.
- C18 spin columns or homemade C18 tips for peptide desalting.

- Incubators with alternating light/dark cycles (12 h) at 21°C–22°C.
- Refrigerated centrifuges (capable of 3500 × *g* to 15,000 × *g*).
- Ultracentrifuge (capable of 100,000 × *g*).
- Ultracentrifuge tubes.
- Mechanical homogenizer (e.g., Minilys) with ceramic tubes and 1.4 mm zirconium oxide beads
- Vortex mixer.
- Mortar and pestle (for grinding with liquid nitrogen).
- Lyophilizer.
- Vacuum concentrator.
- Sonicator (30 kHz, 50 W, adjustable amplitude, cycle, and time).
- EASY-nLC 1200 or equivalent nano-LC system.
- Q Exactive Plus mass spectrometer (Thermo Fisher) or equivalent.
- 16 cm fritless silica columns, 75 μm ID, packed in-house with C18 resin.
- Cooling equipment for solutions and samples (ice, cold blocks).
- Nanodrop 2000 (for protein/peptide quantification).
- Qubit fluorometer (for protein quantification).

2.3 Required Reagents

- 0.01% Tween 20 solution: Prepare a 0.01% (v/v) Tween 20 solution by mixing Tween 20 in water, then sterilize by autoclaving.
  0.1 M ammonium acetate in methanol
- 10% (v/v) glycerol solution: Prepare a 10% (v/v) glycerol solution by mixing glycerol in water, then sterilize by autoclaving.
- 10% TCA in acetone: 10% (w/v) trichloroacetic acid in pure acetone (**Notes 6 and** 7).
- 20% formic acid in water: 20% (v/v) formic acid in ultrapure water.
- 20% TCA in water: 20% (w/v) trichloroacetic acid in ultrapure water.
- 2% DOC solution: 2% (w/v) dodecyl sulfate dissolved in ultrapure water.
- 80% acetone in water: 80% (v/v) acetone in ultrapure water.
- Dense SDS buffer: 30% sucrose, 2% SDS, 100 mM Tris-HCl pH 8.0, 5% β-mercaptoethanol.
- Denaturing buffer: 8 M urea, 100 mM Tris-HCl pH 8.5, and 5 mM DTT.

- Denaturing buffer (Protocol 3.7, phosphoproteome analysis): 6 M urea, 10 mM DTT, in 0.2 M ammonium bicarbonate buffer.
- Denaturing buffer (Protocol 3.8, extracellular vesicles): 8 M urea, 5 mM DTT, in 100 mM Tris-HCl pH 8.5.
- Dilution solution (Protocol 3.7, phosphoproteome analysis): 0.2 M ammonium bicarbonate.
- Elution buffer (Protocol 3.7, phosphoproteome analysis): 5% (v/v) ammonia solution (NH_3) in ultrapure water.
- Loading buffer (Protocol 3.7, phosphoproteome analysis): 80% acetonitrile (ACN) and 6% TFA in ultrapure water.
- Lysis buffer (Protocol 3.4, single-step precipitation): 1% (w/v) SDS in 1× phosphate buffer (0.137 M NaCl, 0.0027 M KCl, 0.01 M Na_2HPO_4, 0.0018 M KH_2PO_4), pH 7.0.
- Minimal salt medium (MSM) + 1% deproteinized tomato cell walls (TCW): Minimal salt medium (50 mM NH_4Cl, 7.3 mM KH_2PO_4, 4.2 mM $MgSO_4$, 6.7 mM KCl, 0.07 mM $FeSO_4$) supplemented with 1% (w/v) TCW and autoclaved.
- Minimal salt medium (MSM) + 1% glucose (GLU): Minimal salt medium (50 mM NH_4Cl, 7.3 mM KH_2PO_4, 4.2 mM $MgSO_4$, 6.7 mM KCl, 0.07 mM $FeSO_4$) supplemented with 1% (w/v) glucose and autoclaved.
- Petri dishes containing malt agar (MA): 2% malt extract and 2% agar (%, w/v).
- Phosphate-buffered saline with 30% sucrose (PBS + 30% sucrose): 0.137 M NaCl, 0.0027 M KCl, 0.01 M Na_2HPO_4, 0.0018 M KH_2PO_4, and 30% (w/v) sucrose in ultrapure water, pH 7.4.
- Polyethersulfone (PES) or cellulose membrane filters: 0.45 μm pore size and 100 kDa molecular weight cut-off.
- Solubilization buffer (Protocol 3.3, classical protein extraction): 8 M urea, 2% (w/v) CHAPS, 20 mM DTT, 0.5% (v/v) Biolytes in 0.1 M Tris-HCl, pH 8.0.
- Solubilization buffer (Protocol 3.7, phosphoproteome analysis): 6 M urea in 0.2 M ammonium bicarbonate buffer.
- TiO_2 micro-columns.
- TFA (trifluoroacetic acid).
- Tris-buffered phenol, pH 8.0 (Biomol, Hamburg, Germany).
- UA buffer (Protocol 3.8, extracellular vesicles): 8 M urea in 100 mM Tris-HCl pH 8.5.

3 Methods

3.1 B. cinerea Preparation and Storage of Conidial Stock

- Preparation and maintenance of conidial stock suspensions were performed in Petri dishes containing malt agar (MA) (2% malt extract, 2% agar (%, w/v)) and maintained in incubators under alternating (12 h) light/dark cycles at 21 °C.
- For storage, conidia were harvested in 0.01% (v/v) Tween 20 solution, filtered through a 30 μm nylon filter, and centrifuged for 5 min at 5000 × *g*. The resulting conidial pellet was suspended to a concentration of 10^7 conidia/mL in a 10% (v/v) glycerol solution for storage at −80°C.
- 0.1% FA (or TFA) + 0.02% DDM: 0.1% (v/v) Formic acid (FA) or trifluoroacetic acid (TFA), and 0.02% (w/v) of n-dodecyl-β -D-maltoside (DDM) in LC-MS grade water.
- LC solvent A: 0.1% (v/v) FA in LC-MS grade water.
- LC solvent B: 80% (v/v) acetonitrile (CAN) and 0.1% (v/v) FA in LC-MS grade water.

3.2 B. cinerea Culture Conditions for Mycelia and Supernatant Sampling Under Virulence Induction

1. For virulence induction, culture media were prepared using two different carbon sources: glucose (GLU) as the constitutive stage; and deproteinized tomato cell walls (TCW) as the virulence inductor.
2. One-liter flasks containing 400 mL of minimal salt medium (MSM) (50 mM NH4Cl, 7.3 mM KH2PO4, 4.2 mM MgSO4, 6.7 mM KCl, 0.07 mM FeSO4) supplemented with 1% of each carbon source were inoculated with fresh 5 × 10^4 conidia/mL of *B. cinerea*.
3. After 5 days of culture at 180 rpm at 22°C under alternating 12-h light/dark cycles, cultures were supplemented with 100 μL of cOmplete™, Mini, EDTA-free Protease Inhibitor Cocktail (Roche) dissolved in 1 mL of 1 M phosphate buffer, pH 7.0. Fungal mycelium was harvested by filtration with a 30 μm nylon filter, allowing separation of the mycelial biomass from the culture supernatant.

3.3 Classical Protein Extraction from Lyophilized Mycelia of B. cinerea

This classical protocol was used to produce the first 2-DE maps [5], serving as an initial point for *B. cinerea* protein extraction.

1. Lyophilize harvested mycelia.
2. Reduce the mycelium to a powder with a pestle and the mortar using N_2.

3. Add 10 mL of 100% acetone (ultrapure and cooled at 4°C) to 2 g of mycelium dust in the bottom of a Falcon 15 mL tube to precipitate proteins.
4. Vortex the sample until it is homogenized, but avoid heating the sample.
5. Centrifuge at 10,000 × *g* for 5 min at 4°C.
6. Repeat steps 3–5.
7. Precipitate proteins from the pellet sequentially with 5 mL of the following solutions (prechilled at 4°C), centrifuging at 10,000 × *g* for 5 min at 4°C between each washing step:
   a. 10% TCA in acetone (repeat this step twice).
   b. 20% TCA in water.
   c. 80% acetone in water (repeat this step twice).
8. Allow the pellet to dry at room temperature.
9. Resuspend the pellet in 5 mL of "Dense SDS buffer" by vortexing.
10. Add 5 mL Tris-buffered phenol pH 8 and mix by vortexing for 30 s.
11. Centrifuge 10,000 × *g* for 10 min at 4°C.
12. Collect the phenolic top phase in COREX tubes.
13. Add five volumes of 4°C prechilled 0.1 M ammonium acetate in methanol to the phenolic phase to precipitate proteins.
14. Incubate overnight at −20°C.
15. Centrifuge at 10,000 × *g* for 10 min at 4°C
16. Wash the pellet by precipitating proteins sequentially with: a. Cold 0.1 M ammonium acetate in methanol (twice). b. Cold 80% (v/v) acetone (twice).
17. After drying the pellet, resuspend it in the solubilization buffer for further analysis.
18. Follow 2-DE protein separation or a protein digestion protocol joined to a peptide desalting protocol for LC-MS/MS analysis desalting protocol for LC-MS/MS analysis (**Note 8**).

3.4 Single-Step Precipitation Protocol for Protein Extraction from *B. cinerea* Mycelia for LC-MS/MS

At present, the improved method based on specific matrices and homogenization instruments, joined to simplification of the protein precipitation protocol, has improved the ratio of obtained proteins with shorter protocols while maintaining the quality of the extracts [6] (**Notes 9 and 10**).

1. Extract lyophilized mycelia by grinding 0.3 g of mycelia in 6 mL of lysis buffer (PBS +1% SDS), using a homogenizing kit with ceramic beads (1.4 mm zirconium oxide beads) in a mechanical

homogenizer Minilys (Bertin Technologies) at maximum level, five cycles of 30 s.

2. Clarify lysates by centrifugation at 10,000 × *g* for 10 min at 4°C, retaining the supernatant.
3. Precipitate proteins by adding six volumes of cold acetone and incubating overnight at −80°C.
4. Precipitate protein pellet by centrifugation at 10,000 × *g* for 15 min at 4°C.
5. Dry pellets may be resuspended in the denaturing buffer.
6. Follow 2-DE protein separation and digestion protocol, or directly a protein digestion protocol joined to a peptide desalting protocol for LC-MS/MS analysis (**Notes 11-13**).

3.5 Secretome Extraction

The secretome is one of the most relevant subproteomes, especially in fungi. Here, we present the protocol used for the first description of this subset of proteins in *B. cinerea* [7].

1. Separate fungal mycelia from the liquid media by filtration using a 30 μm nylon filter.
2. Then centrifuge supernatant at 15,000 × *g* for 30 min at 4°C to remove cellular debris.
3. Add protease inhibition cocktail.
4. Divide the supernatant into 30 mL aliquots.
5. Add 1/100 (v/v) of 2% DOC to the sample and incubate on ice for 30 min.
6. Add 1/10 (w/v) of cold 100% TCA to the sample and vortex immediately to avoid the formation of clumps that can trap contaminants.
7. Let the sample precipitate for 1 h on ice. Preferably overnight at −20°C.
8. Centrifuge at 10,000 × *g* for 10 min and remove the supernatant.
9. Add 10 mL of cold 80% acetone to the protein pellet and centrifuge it at 10,000 × *g* for 10 min at 4°C.
10. Repeat step 9 twice.
11. Air-dry the pellet.
12. The dried pellet may be resuspended in the required buffer or in Dense SDS for phenol precipitation (see steps 10–17 of Sect. 3.3)
13. Follow 2-DE protein separation and digestion protocol, or directly a protein digestion protocol joined to a peptide desalting protocol for LC-MS/MS analysis.

3.6 Extraction of Surface Proteins (Surfactome)

Protein signaling pathways are the key elements that transmit information from environmental conditions to the inner cytoplasm to construct an optimal response to complete their life cycle, with surface proteins serving as the front-line receptors [8] (Fig. 2).

1. Separate the mycelium by centrifugation at 5000 × *g* for 5 min from 10 mL of culture.
2. Wash the mycelium three times with 1 mL of PBS with 30% sucrose (pH 7.4).
3. Add 10 μg trypsin to 1 mL PBS with 30% sucrose (pH 7.4) and incubate at 37°C for 5 min.
4. Centrifuge at 5000 × *g* for 10 min and discard the pellet.
5. Filter supernatant with 0.22 filters.
6. Incubate overnight at 37°C.
7. Add TFA at a final concentration of 0.1% to stop the reaction. Incubate on ice for 15 min.
8. Desalt peptides using C18 spin columns for LC-MS/MS analysis.

3.7 Phosphoproteome Analysis in B. cinerea

One of the most relevant protein posttranslational modifications (PTM) is based on the use of phosphate groups to modify protein structure and function. This PTM serves as a key instrument in the translation of signal cascades. These kinds of modifications may only be studied through proteomics. In an attempt to describe the relation between phosphate modifications of proteins and the regulation of the infection cycle, we need specific protocols for *B. cinerea* phosphopeptide enrichment [9].

1. Start with a total extract of proteins obtained, for example, using a previously reported phenol-based procedure.
2. Dissolve the protein pellet in 300 μL of the solubilization buffer (6 M urea, 0.2 M ammonium bicarbonate).

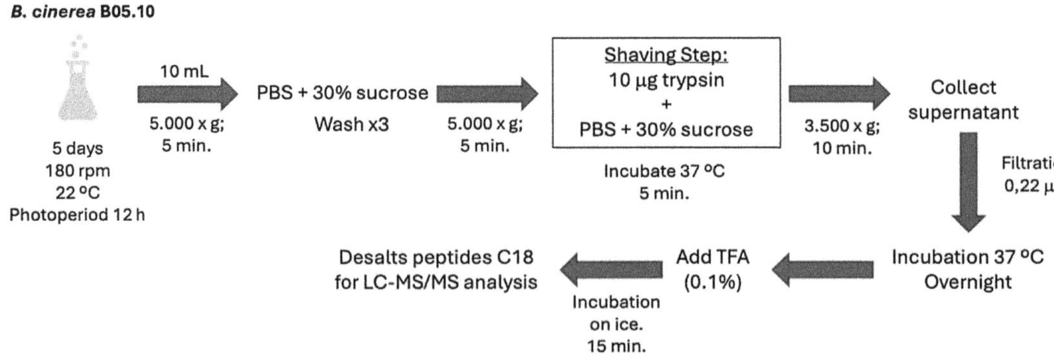

Fig. 2 Schematic description of the process to obtain the "surfactome" from *B. cinerea* cultures

3. Protein digestion:
   a. Reduce 500 µg of each sample by dissolving them in denaturing buffer (6 M urea, 10 mM DTT, 0.2 M ammonium bicarbonate) and incubating at 37°C for 60 min.
   b. Alkylate the protein by adding iodoacetamide to a final concentration of 20 mM and incubating at 25°C for 30 min.
   c. Dilute samples up to 2 M urea with dilution solution (0.2 M ammonium bicarbonate) and digest them overnight with Lys-C at 37°C.
   d. Dilute again twofold prior with dilution solution and digest overnight with trypsin at 37°C.
   e. Desalt crude peptide mixtures using a C18 column.
4. Phosphopeptide enrichment:
   a. Prepare TiO_2 micro-columns in gel loading tips (0.5 mg) and equilibrate them with the loading buffer (80% ACN in water + 6% TFA)
   b. Enrich with phosphopeptide 250 µg of each digested sample using TiO_2 micro-columns.
   c. Load samples in the loading buffer and wash the columns with 80% ACN in water + 0.1% TFA
   d. Elute phosphopeptides with 30 µL of elution buffer (5% NH3 in water) into a 1.5-mL tube containing 30 µL of 20% formic acid in water.
5. Desalt crude phosphopeptide mixtures using a C18 column.

3.8 *Botrytis cinerea* Extracellular Vesicles Proteome

Extracellular vesicles are the cytosolic elements involved in the transportation of biological macromolecules. This trafficking of compounds seems to direct fungal host interactions and cell communication [6]. Here we present the technical approach to studying their protein content (Fig. 3):

1. Supplement 400 mL of 5-day-old *B. cinerea* cultures with Protease Inhibitor Cocktail (Roche).
2. Remove *B. cinerea* mycelium from the culture medium by filtration with a 30 µm nylon filter.
3. Remove cell debris and spores from broth cultures by centrifugation at 4°C for 25 min at 4500 × *g* and for 45 min at 15,000 × *g*.
4. Filter the supernatant using a polyethersulfone (PES) or cellulose membrane, 0.45 µm cutoff filter.
5. Concentrate the supernatant with an ultrafiltration PES or cellulose filter (cutoff filter 100 kDa).

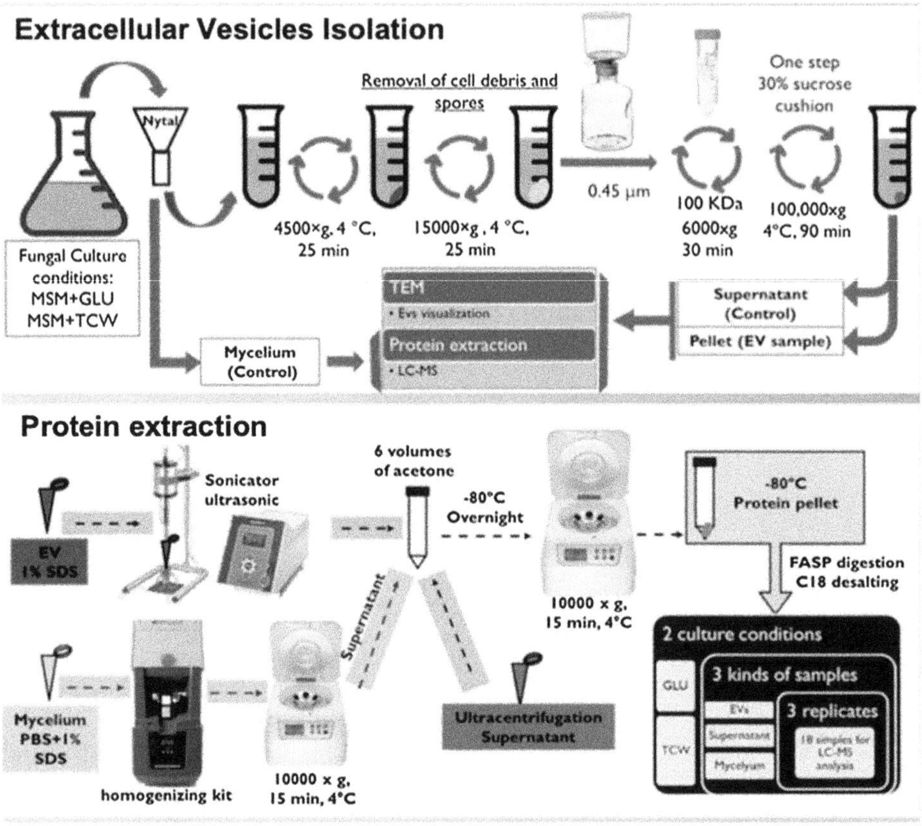

Fig. 3 Workflow and brief description of the protocol to prepare proteomics analysis of *B. cinerea* vesicles

6. Load 30 mL of vesicle-enriched supernatant slowly over 4 mL of 1× PBS + 30% sucrose solution and centrifuge it at 100,000 × *g* and 4°C for 90 min.
7. Separate the top supernatant from the bottom sucrose layer (~5 mL).
8. Centrifuge the top supernatant for 80 min at 100,000 × *g* and 4°C to pellet the remaining EVs.
9. Dilute this EV pellet with 10 ml of 1× PBS, and load it onto the previously obtained EV-sucrose layer and ultracentrifuge it for 90 min at 100,000 × *g* and 4°C to pellet the EVs.
10. Remove the top supernatant completely, and retain the EV-containing sucrose layer.
11. Dilute the EV-containing sucrose layer with 30 mL of 1× PBS, pH 7.4 and then ultracentrifuge it at 100,000 × *g* for 90 min at 4°C.
12. Wash the EV pellet with 10 mL of 1× PBS, pH 7.4, by ultracentrifugation at 100,000 × *g* for 90 min at 4°C.

13. Resuspend the EV pellet in 1 mL of sterile 1× PBS (pH 7.4). Protein content was measured, and EV samples were stored at −80°C.
14. EV protein extraction:

    a. Supplement the EV aliquots with 1% SDS and sonicate them twice at amplitude 80%, cycle 0.5, 50 W, 30 kHz, for 30 s on ice.
    b. Precipitate proteins from vesicle suspensions by adding six volumes of cold acetone and incubating overnight at −80°C.
    c. Precipitate proteins by centrifugation at 10,000× *g* for 15 min at 4°C.

15. Protein digestion using the FASP protocol:

    d. Dissolve the protein pellet in denaturing buffer (8 M urea, 100 mM Tris-HCl pH 8.5, 5 mM DTT) and incubatefor 30 min at room temperature.
    e. Protein aliquots in buffer UA (8 M urea, 100 mM Tris-HCl pH 8.5) were added to a 30 kDa cutoff filter unit and centrifuged at 14,000× *g* for 10 min.
    f. Wash the filter with 450 μL of buffer UA at 14,000 × *g* for 10 min.
    g. Alkylation of the protein was performed by adding 100 μL of the CAA solution (55 mM chloroacetamide, 8 M urea, 100 mM Tris-HCl pH 8.5) and incubating at room temperature for 30 min in the dark.
    h. After centrifuging the filter at 14,000 × *g* for 10 min and washing it twice with 450 μL of buffer UA, 450 μL of dilution solution (100 mM Tris-HCl, pH 8.5, 1 mM CaCl2) was added.
    i. For protein digestion, a 1:100 trypsin-to-protein ratio is applied at 37°C overnight.
    j. Peptides were collected by centrifuging at 14,000 × *g* for 10 min and acidified with 20% TFA (trifluoroacetic acid) solution (final concentration 0.5%).
    k. The peptide solution was cleaned using a C18 matrix, equilibrated with 75 μL of methanol and 75 μL of solution B (80% acetonitrile, 0.5% FA), and 75 μL of solution A (2% acetonitrile, 0.1% TFA) by centrifugation at 1200 × *g* for 3 min.
    l. The peptide solution was loaded by centrifugation at 800 × *g* for 5–10 min. Peptides were eluted with 25 μL solution

B by centrifugation at 500 × *g* for 5 min. The eluted peptide was dried in a vacuum concentrator.

m. The peptide pellet was resuspended in 10 μL buffer A* (2% acetonitrile, 0.1% TFA), and peptide concentration was measured in a Nanodrop 2000.

n. The peptide solution was diluted to 0.1 μg/μL for LC-MS analysis.

3.9 Proteomic Analysis by LC-MS of *Botrytis cinerea*

As the last step of each proteomics approach, the analysis of the obtained peptides by LC-MS shows the success of the whole process, mainly based on the number of identified proteins [4]. These procedures may be optimized depending on your specific experimental design, the instrument used, the size of the database, etc. For this reason, this protocol must be used only as a starting point.

1. Extracted proteins must always be digested (in-gel digestion, FASP digestion, or in-solution digestion) and desalted using C18 columns.
2. Dried peptides were re-dissolved in10 μL of 0.1% FA (or TFA) + 0.02% DDM.
3. Dilute the sample with 0.1% FA (or TFA) + 0.02% DDM to 0.1 μg/μL for analysis.
4. Analyze samples using an EASY-nLC 1200 coupled with a Q Exactive Plus mass spectrometer (Thermo Fisher).
5. Separate peptides using 16 cm frit-less silica emitters (75 μm inner diameter), packed in-house with reversed-phase ReproSil-Pur C18 AQ 1.9 μm resin.
6. Elute peptides from the column for 115 min using a segmented linear gradient of 5%–95% solvent B (0–5 min: 5% of B; 5–65 min from 5% to 20% of B; 65–90 min: from 20% to 35% of B; 90–100 min: from 35% to 55% of B; 100–105 min: from 55% to 95% of B; 105–115 min: 95% of B) (solvent A 0% ACN, 0.1% FA; solvent B 80% ACN, 0.1% FA) at a flow rate of 300 nL/min.
7. Acquire mass spectra in data-dependent acquisition mode using a TOP15 method. MS spectra were acquired in the Orbitrap analyzer with a mass range of 300–1750 m/z at a resolution of 70,000 FWHM and a target value of 3 × 10^6 ions.
8. Select precursors with an isolation window of 1.3 m/z.
9. Perform HCD fragmentation at a normalized collision energy of 25.
10. Acquire spectra with a target value of 10^5 ions at a resolution of 17,500 FWHM, a maximum injection time of 55 ms, and a fixed first mass of m/z 100.

11. Peptides with a charge of +1, greater than 6, or with an unassigned charge state must be excluded from fragmentation for MS2. Dynamic exclusion for 30 s prevents repeated selection of precursors.
12. Process obtained raw data using specific software such as MaxQuant and Perseus, or Proteome Discoverer.

4 Notes

1. Proteomics techniques have several contaminants that may disturb the process; chemical contaminants can be removed during the extraction protocols. However, the contamination of exogenous proteins, such as keratin, must be avoided during manipulation. Contact with skin, hair, silk, and dust must be avoided; therefore, the use of nitrile gloves and lab clothes is mandatory.
2. The described protocols assume the use of organic solvents, some of which can degrade the polymers of the plastic tubes. These residues can disrupt the identification process, but they can also degrade the chromatography matrices used. High levels of plastic tubes must be used.
3. The protocols described in this chapter are not written as fixed and unchangeable paths; rather, they are a set of methodologies that should be considered, taking the parts that interest us most depending on our own experimental design.
4. Protein and peptide quantification may be performed by several methods; we primarily used Nanodrop and Qubit instruments.
5. Add to the dilution buffer any cofactors or additives required by each digestion enzyme, following the manufacturer's instructions.
6. When performing TCA precipitation, strictly control the incubation time, as prolonged exposure may affect protein recovery and integrity.
7. During acetone precipitation, proteins can be stored for an extended period at this step before proceeding, allowing flexibility in workflow.
8. Peptide desalting with C18 columns is an essential step prior to LC-MS/MS analysis. Several formats are available, including spin columns or homemade C18 tips.
9. Use low-binding materials to prevent protein and peptide loss.

10. For proteomics, preferably use plastics such as polyvinyl (PV) or polypropylene (PP). The use of polyethylene glycol (PEG) is strictly prohibited to prevent protein and peptide contamination.
11. Always use LC-MS grade water for LC-MS solutions. For other dilutions and resuspensions, use ultrapure water to avoid contaminants that could interfere with analysis. Whenever possible, it is recommended to use LC-MS grade water for all steps.
12. Protein digestion in-gel or via FASP protocols is preferred, as these methods effectively remove salts and other contaminants that can interfere with analysis, providing higher efficiency and reproducibility for LC-MS sample preparation.
13. Perform all protein extraction steps using cold solutions and keep samples on ice whenever necessary, avoiding heating during extraction to prevent protein degradation. The use of protease inhibitor cocktails is also strongly recommended.

References

1. Fernández-Acero FJ, Jorge I, Calvo E, Vallejo I, Carbú M, Camafeita E, López JA, Cantoral JM, Jorrín J (2006) Two-dimensional electrophoresis protein profile of the phytopathogenic fungus Botrytis cinerea. Proteomics 6(Suppl 1):S88–S96. https://doi.org/10.1002/pmic.200500436
2. Fernández-Acero FJ, Jorge I, Calvo E, Vallejo I, Carbú M, Camafeita E, Garrido C, López JA, Jorrín J, Cantoral JM (2007) Proteomic analysis of the phytopathogenic fungus Botrytis cinerea as a potential tool for identifying pathogenicity factors, therapeutic targets, and for basic research. Arch Microbiol 187(3):207–215. https://doi.org/10.1007/s00203-006-0188-3
3. Fernandez-Acero F, Carbu M, Garrido C, Vallejo I, Cantoral J (2007) Proteomic advances in phytopathogenic fungi. Curr Proteomics 4:79–88. https://doi.org/10.2174/157016407782194620
4. Coste OBN, Escobar-Niño A, Fernández-Acero FJ (2025) Proteomics-based approaches to decipher the molecular strategies of *Botrytis cinerea*: a review. J. Fungi 11:584. https://doi.org/10.3390/jof11080584
5. Fernández-Acero FJ, Colby T, Harzen A, Cantoral JM, Schmidt J (2009) Proteomic analysis of the phytopathogenic fungus Botrytis cinerea during cellulose degradation. Proteomics 9(10):2892–2902. https://doi.org/10.1002/pmic.200800540
6. Escobar-Niño A, Harzen A, Stolze SC, Nakagami H, Fernández-Acero FJ (2023) The adaptation of Botrytis cinerea extracellular vesicles proteome to surrounding conditions: revealing new tools for its infection process. J Fungi 9:872. https://doi.org/10.3390/jof9090872
7. Fernández-Acero FJ, Colby T, Harzen A, Carbú M, Wieneke U, Cantoral JM, Schmidt J (2010) 2-DE proteomic approach to the *Botrytis cinerea* secretome induced with different carbon sources and plant-based elicitors. Proteomics 10:2270–2280. https://doi.org/10.1002/pmic.200900408
8. Escobar-Niño A, Carrasco-Reinado R, Morano IM, Cantoral JM, Fernandez-Acero FJ (2021) Unravelling the initial triggers of Botrytis cinerea infection: first description of its surfactome. J Fungi (Basel) 7(12):1021. https://doi.org/10.3390/jof7121021
9. Liñeiro E, Chiva C, Cantoral JM, Sabido E, Fernández-Acero FJ (2016) Phosphoproteome analysis of *B. cinerea* in response to different plant-based elicitors. J Proteomics 139:84–94. https://doi.org/10.1016/j.jprot.2016.03.019

Index

Luca Nerva and Walter Chitarra (eds.), *Botrytis cinerea*, Methods in Molecular Biology, vol. 3076,
https://doi.org/10.1007/978-1-0716-5551-1,

F

G

H

I

O

P

R

S

T

U

V

GPSR Compliance
The European Union's (EU) General Product Safety Regulation (GPSR) is a set of rules that requires consumer products to be safe and our obligations to ensure this.

If you have any concerns about our products, you can contact us on

ProductSafety@springernature.com

In case Publisher is established outside the EU, the EU authorized representative is:

Springer Nature Customer Service Center GmbH
Europaplatz 3
69115 Heidelberg, Germany

www.ingramcontent.com/pod-product-compliance
Ingram Content Group UK Ltd.
Pitfield, Milton Keynes, MK11 3LW, UK
UKHW052041290926
476116UK00005B/154
* 9 7 8 1 0 7 1 6 5 5 5 0 4 *